Red Flag Warning

Red Flag Warning

Mutual Aid and Survival in California's Fire Country

Edited by
Dani Burlison and Margaret Elysia Garcia

Introduction by
Manjula Martin

Red Flag Warning: Mutual Aid and Survival in California's Fire Country
Dani Burlison and Margaret Elysia Garcia © 2025
This edition © 2025 AK Press
All essays © 2025 by their respective authors, except where otherwise noted

ISBN 9781849356084
E-ISBN 9781849356091
Library of Congress Number: 2024949077

AK Press
370 Ryan Avenue #100
Chico, CA 95973
www.akpress.org
akpress@akpress.org
510.208.1700

AK Press
33 Tower St.
Edinburgh EH6 7BN
Scotland
www.akuk.com
akuk@akpress.org

Excerpt(s) from *The Last Fire Season: A Personal and Pyronatural History* by Manjula Martin, copyright © 2024 by Manjula Martin. Used by permission of Pantheon Books, an imprint of the Knopf Doubleday Publishing Group, a division of Penguin Random House LLC. All rights reserved.
Cover artwork by Saif Azzuz
Cover design by Crisis
Printed in the USA on acid-free paper

Contents

Preface vii

Introduction: A Brief Pyronatural History of California
by Manjula Martin 1

Solidarity, Not Charity: Mutual Aid in Chico, California
by Hiya Swanhuyser 9

UndocuFund: Grassroots Solidarity Networks of Mutual Aid for Liberated Futures
by Beatrice Camacho 19

The Community of Evacuation
by Margaret Elysia Garcia 26

What Wildfires Do to Our Minds
by Dani Burlison 41

Heaven over Fire: Finding Community in a Burn Scar
by Jane Braxton Little 51

Formerly Incarcerated Firefighters as Community Servants: An Interview with Brandon Smith of Forestry and Fire Recruitment Program
by Dani Burlison 73

Listening to the Loved One: Wildfire, Community, and Ambiguous Loss
by Amy Elizabeth Robinson 85

Parenting in Fire Country: An Interview with Kailea Loften
by Dani Burlison 101

Funding and Finances During and After Disaster
by Sue Weber 113

Ground Truth: The Limits of Scale
by Zeke Lunder 119

Triptych in Smoke
    by Lasara Firefox Allen 127

Bringing Fire Back to the Land: An Interview with Margo Robbins of Cultural Fire Management Council
    by Dani Burlison 133

The Little Sawmill That Could
    by Margaret Elysia Garcia................................145

We Circle Up for Fire: A Community Burn
    by Redbird Willie... 151

    Afterword..163

    Acknowledgments 167

    Contributors 169

Preface

The idea for this book came to us during a California fire season, the time of year that sometimes starts in May (from-never-quite adequate snowpack or rainfall the winter and spring before) and ends in October or November. Each season we hope we are lucky, and each season more of us are not.

Fire season is always stressful for folks in Northern California, but the fire that really drove home the inevitability of continued major climate catastrophes for one of the editors, Dani, was the October 2017 North Bay Firestorm. The Tubbs Fire, one of numerous fires in the region, burned over 36,000 acres where she lived (and had lived for thirty years). It roared over the mountains just east of Santa Rosa at night, chewing up thousands of homes and burning ruthlessly for three weeks, while the Redwood Complex fire burned to the north, in Mendocino County. Both fires took lives and destroyed thousands of homes.

During the fires, however, Dani witnessed an amazing surge of mutual aid efforts anchoring around Sonoma County: pop-up mental health clinics, fund-raising efforts for undocumented people, holistic healing pop-ups for firefighters and evacuees, DIY evacuation sites on private property for undocumented workers, bioremediation efforts to protect local water sources, language justice in the form of assistance for non-English speakers in shelters, free food, yoga classes, clothing drives, more people eager to learn about Indigenous fire practices, and more. She began writing articles about how folks were navigating home loss around Northern California, how communities were banding together to rebuild and support each other. Growing up in a rural, working-poor community in Tehama County—where the Park Fire burned through 315,000 acres in 2024 as we put finishing touches on this manuscript—Dani watched several special places from her rural childhood burn up in Butte, Shasta, Lassen, and other counties. It was a lot.

Simultaneously, and for the better part of five years, Margaret was a reporter and columnist for Feather Publishing, whose four Plumas County newspapers were known as some of the last remaining independent papers in the West. During COVID-19, the newspapers ceased publication, and regional news was then posted to the website *Plumas News* (www.plumasnews.com). During the Dixie Fire of 2021, news was posted several times a day. Margaret had been evacuated a few times but hadn't felt the totality of fire. Then came *her* fire. It burned over 960,000 acres. She wanted to remember the details even as they were overlooked for more-urban perspectives on fire. She wanted to change the narrative set forth by national news and have people's voices heard, and she wanted to report on hope and community coming together. Margaret wanted to collect stories from the fire that focused on issues like dealing with smoke, evacuations, and resourcefulness in low-income communities. She began a column at *Plumas News* called "Greenville Rising" and then "Communities Rising" as more communities faced the difficult task of rebuilding their lives and communities after fire.

Between Sonoma and Plumas counties, we texted one another about all the tangential elements of wildfire disaster that we weren't hearing, the parts that didn't fit into a neat, quick segment; the parts that weren't for viewers, readers, or listeners far away from the disaster; but rather the stories and information needed for those in the thick of it: the community—the voices of people who had been living rurally for decades and knew and loved these mountains and forests and communities. We laughed about tales of so many firefighters flooding friends' dating apps. We worried about the same firefighters getting enough sleep. We checked in with each other about evacuations, offering each other a safe, fire-free place to stay, relaying stories about community members' heroics, and complaining about outsiders—those with no experience living in rural California—who wrote disparagingly about fire survivors in places outside of Northern California's typical tourist destinations that frequently take so much of the fire spotlight in mainstream media.

It was the lack of rural voices and the stories of mutual aid that led us to create this anthology, something you can hold in your hands with some of Northern California's finest essayists and minds giving voice to living through these giant climate catastrophes.

The contributors to this book are just a small slice of the wide array of people impacted by California's fires and working toward climate and community resiliency. Yet, in our minds, they offer essential perspectives. We definitely can't cover every aspect of living in a fire ecology in this book. While many see fires as great equalizers, and climate anxiety is something felt across class and race lines, the ways fire impacts each individual and each community can differ depending on factors like class, citizenship status, location, geography, population size and demographics, housing status (renting vs. homeowning), and Indigenous knowledge (or lack of it). What we wish we had room to include is discussion of the dire mental health consequences for low-income people, firefighters, immigrants, people with disabilities, seniors, and others affected by wildfires. We also wish we had room to include all of the individuals, mutual aid networks, and nonprofit organizations (like North Bay Jobs with Justice, Sonoma County Legal Aid, and others) who have worked tirelessly to center and support farmworker organizing and legal support around housing for low-wealth families.

Red Flag Warning explores the ways these fires take root and impact rural and urban Northern California. It examines our relationships to place and community. We want to share our love of Northern California and its people and ecosystems. We also want the reader to examine what fire can and does mean to them, what it means to reimagine the world, to prepare for the worst, to examine flames through different lenses. We want readers to understand the importance of mutual aid, of organizing, of community care, of resilience.

For the reader outside of Northern California, we hope this anthology brings more understanding of what we lived through and will continue to live through as our mountainous forests of mixed pines and oak woodlands become mountains with a few new root balls of oak trees and museum pieces of pine, and as more people from California join the growing global population of climate refugees. For the Northern Californian, may this serve as an extension of our complex and diverse communities and provide mutual aid for the soul. We've attempted to address the many aspects of living with fire that we face as we learn more and more to rely on each other for both solace and recovery—whatever that looks like for each of us.

With love and solidarity from California,
Dani & Margaret

Introduction
A Brief Pyronatural History of California
Manjula Martin

When fire first came to the land now called California, it came to stay.[1] It came as lightning, magma, mountains. Fire was always a naturally occurring part of the landscape in the western United States. From forests to grassland to desert, the diverse collection of ecosystems that made up this vast region evolved with fire. The land depended on the cycles of renewal fire brought, and the people who had lived on the land since time immemorial learned to manage this place with an understanding that fire would always need to be a part of it. The relationship between fire, land, and people was interdependent and continually evolving. While precolonial Indigenous land management practices did have a disruptive effect on ecosystems—all human presence did—they did so in a way that achieved balance between individuals and community, people and nonhumans. Fire cleared dangerous fuels from the landscape, preventing more-destructive fires. It helped hunters direct the course of game. Fire encourages the health and frequent regrowth of plants necessary for survival of people and wildlife. The acorns of young oaks provided a staple food; newly resprouted reeds and grasses

---

1. This introduction is partially excerpted from Manjula Martin, *The Last Fire Season* (New York: Pantheon, 2024).

were used to make baskets and boats. The specifics of precolonial Indigenous land management and cultural practices differed between communities, but fire figured prominently in most of them. Fire was and remains a primordial physical and spiritual presence—an element, essential to all life.

When new people came to this land in the 1500s, they called themselves Europeans, Spanish, and later Californios. They kept coming, on and off, for hundreds of years, as did people who called themselves Mexican, Russian, or American. And they brought with them their god. During the Mission Era, from the mid-1700s to the mid-1800s, California's Indigenous people were displaced, forcibly converted, and enslaved under the missionary systems of settlement carried out largely by Catholic priests (*padres*, in Spanish)—systems explicitly designed to remove people from their land and convert them into a productive peasant population. Then somebody struck gold, and the floodgates of settlement opened even wider.

Beginning with the missions, colonialists had recognized the importance of fire to Indigenous life. In 1793 the acting governor of California under Spanish rule, Don José Joaquín de Arrillaga, issued an edict outlawing the use of fire by Native peoples, even in remote areas, in order to preserve pastureland for Spanish cattle concerns. By 1821, possession of California passed from Spain to Mexico. Alta California, as the province was then known, relied increasingly on livestock economies; fire was used by some landowning rancheros to clear pasture land, but aboriginal fire use was still suppressed. In addition to being important to Indigenous people for ecological and cultural reasons, fire was a tool of war and as such might be used in rebellion. Fire was strategically used in Indigenous uprisings against colonialization in 1776 at Mission San Luis Obispo, when rebels set fire to the roofs of the compound, and in the 1824 Chumash revolt, which led to the occupation of three Southern California missions for several months before being violently quashed by Mexican military forces. Two years later, in what is now Sonoma County, a Native resistance movement—likely including Wappo, Coast Miwok, Suisin, and Patwin peoples—used fire to burn several buildings in the

Mission San Francisco Solano there, ultimately ousting the violently cruel *padre* in charge.

In 1850, California became a US state after being bounced between colonizers for hundreds of years. One of the first things the California legislature did was create the ironically titled 1850 Act for the Government and Protection of Indians, also known as the Indian Indenture Act. The law prohibited the use of intentional fire in the new state. It also effectively legalized the kidnapping and enslavement of Native people, including the sex trafficking of girls and women. This law and many others like it were part of a coordinated campaign to *exterminate*—that was the word the then-governor used—the Indigenous population of the new state. The campaign was so blatant that most local governments offered cash bounties to settlers for the body parts of Native people, as proof of their murder. The Indian Indenture Act was revised in 1860 to expand the allowance of enforced labor, permitting Native children to remain enslaved well into adulthood. Over the next twenty years, federal, state, and local governments organized or enabled the killing of 80 percent of the Indigenous people then living in California, an act of genocide that, among other things, erased or pushed underground millennia of traditional ecological knowledge, including the culture of fire and related practices. Despite the prohibitions, Indigenous survivors of the genocide continued to use fire on the land when they were able, as would their descendants.

In the late 1800s, conservationists fell in love with California's epic landscapes and sought to preserve them for the use of the public, by which they meant mostly people like themselves. It was assumed or decided that the well-tended forests in places like the Sequoia Forest or the Yosemite Valley were naturally occurring and not the result of careful land management over millennia. This view of nature—as a thing separate from humans, a museum of the outdoors with no occupants and frozen in time—combined with the lie that Indigenous people were unsophisticated in their ways of living, thus cementing the American mythology of the West as a place that was both wild and virginal.

Settler culture and old-school conservation culture met in a fear of fire. Beginning in the 1800s, some US forestry professionals, usually those with firsthand field experience, advocated for forest management policies that emulated long-standing aboriginal practices. Because of a timeless combination of racism, profit motives, and the prioritization of recreation and housing over the health of the landscape, they lost. The US Forest Service was established in 1905, with fire prevention and suppression as primary goals. Total fire exclusion was further codified after the Great Fire of 1910, nicknamed the Big Burn or the Big Blowup, a wildfire driven by extreme winds that killed eighty-nine people and torched three million acres in Montana, Washington, and Idaho over the course of two days. The Big Burn further swayed public opinion against the idea that a fire might be capable of doing anything but causing disaster. By 1935, the Forest Service had instituted the 10:00 a.m. policy, which stated that every new wildfire needed to be extinguished by ten o'clock the next morning. The policy was in place until the mid-1970s. In practice, some non-Indigenous landowners in California were able to use prescribed burning without legal consequence well into the 1940s and 1950s, until popular opinion turned with the help of Smokey Bear, the Forest Service's threateningly peppy anti-fire mascot. It wasn't until the late 1960s that some forestry management entities began to officially reintroduce the practice of controlled burning. By then, the seeds of the current cataclysm had already germinated.

Since the days of colonization and extraction, a deadly mix of factors created the wildfire crisis. The genocide led to the erasure and criminalization of traditional ecological knowledge and practices. It also, in combination with settler conservation philosophies, enforced a culture of fear around fire, further evangelizing the ideology that nature was outside of and subject to humanity. Total fire exclusion and rampant resource extraction led to overcrowded and unhealthy forests that, without regular fire regimes, encroached on open grasslands, creating more heavy fuel. Meanwhile, California's perpetual cycles of boom-and-bust development and housing crises continued to send people of all economic classes into the wildland-urban interface (known as the WUI and jarringly pronounced *woo-ee*), where built environments touched up against so-called wild places. (I was one of those people.) The increased human presence in the WUI primed landscapes with even more fuel, in the form of houses and businesses. And the big one: human-caused climate change from the burning of fossil fuels caused an increase in extreme weather, pushing California's diverse ecosystems—which were already adapted to regular cycles of drought, flood, and fire—toward new polarities. Winter weather oscillated more wildly between record drought and record rainfall. More rain made more vegetation, while hotter dry seasons transformed all that moist, vigorous plant growth into dry kindling. Abundant fuel stoked bigger, hotter, and faster fires. At the same time, increasingly severe winds helped turn regularly occurring wildfires into megafires; more destructive wildfires further decimated ecosystems and property. It was a perfect storm of causation.

California's ecosystems—like those in many parts of the Americas, the Mediterranean, and Australia, to name just a few—are fire-adapted. The land here has been accustomed to regular fire, both wild and human-tended, and hasn't been getting it. But even in a fire-adapted landscape, not all fire is good. These new extreme, drought- and wind-fueled wildfires have the capability to permanently alter ecosystems. Fire too is changing. Wildfire is now occurring in environments entirely unaccustomed to it, such as Siberia, Sweden, the British Isles, and Hawaii. The wildfire crisis, like climate change, is global. The megafire future is now.

· · · · ·

It can be difficult to quantify the widespread impact of this pyro-natural history. Each new set of numbers detailing the impact of the current crisis are no sooner tallied than they are exceeded: acres burned, structures destroyed, lives lost. Species displaced. Beloved landscapes rendered unrecognizable. To those of us living in the places that fire touches, it can sometimes feel like everything is unprecedented, every time. The names of past wildfires become as familiar as street names or memories from childhood—imprinted, seemingly monolithic—yet at some point they all blur together. Camp, Carr, Dixie, Tubbs. Human actions affect climate change, extreme weather disasters affect humans, and the cycle grows and boomerangs and accelerates until climate catastrophe is indecipherable from everyday life. And, as always, disasters are intricately bound up with societal failures: it's the most at-risk populations who are hit hardest. What people living with fire have learned through experience is that there are other ways to count.

"Red flag warning" is a term used by California fire and weather authorities to categorize fire risk. In my memory, red flag warnings primarily occurred during the dry season in the West, commonly known as "fire season." But as harmful, unhealthy wildfires increase, and the perfect storm intensifies, fire season is no longer limited to specific months. The term more commonly used by Cal Fire (the state's Department of Forestry and Fire Protection) is "fire year," but the phrase that might better describe my and other Californians' experience of wildfire is one that is becoming common in ecological and emergency management circles: "living with fire." Beyond its literal meaning, the prospect of living with fire necessitates the deeper acknowledgment that wildfire is inseparable from other aspects of life on Earth.

Written by Californians who have been directly impacted by life with fire, this book collects stories of experience, loss, action, and shared resources and knowledge. The voices gathered here speak from many different vantage points and include those of Indigenous firekeepers leading the movement to restore good fire, those in small towns whose social fabric has been rewoven stronger by fire, and those providing game-changing examples of organizing and mutual aid from the fields to the shelters of Northern California. The focus is on Northern California, but it's our hope that communities elsewhere facing their own types of climate consequences, whether in the form of fire, water, or infrastructure decline, can find much to relate to in this collection. Whether we live in a place of fire or ice, in hurricane country or tornado alley, people everywhere are responding to the crisis of climate disaster with a resounding commitment to solidarity. It's not enough to merely survive. Our lives, relationships, and ecologies demand more. The communities represented in this book are bravely, beautifully leaving behind the false sense of security that so much of us rely on to navigate quotidian life during climate change—what writer Barry Lopez called "unwarranted hope"—and seeking new ways of living with fire.

In many ways, *Red Flag Warning* arrives as less a warning and more of a reminder, a clarion call of a simple, revolutionary truth that can no longer be ignored: people are a part of the natural world. It has never been more urgent to offer respect and care to each other and the land, and in doing so to lift up the leadership of Indigenous communities in finding ways forward.

As the history of people and fire shows, acts of mutual aid are not limited to humankind. There is an eternal exchange of care between humans and the elements, and the act of tending fire and land is also a form of mutual aid. Many people across the American West and the world are already challenging typical disaster narratives and looking to fire itself—good fire, cultural fire, spiritual fire, fire a teacher and as a peer—for solutions. Humanity has evolved in relationship to fire, and we shall continue to do so. The only question is how. Past and current-day acts of capitalist colonialism, with their infinite iterations of extraction and harm, have led humanity to this moment. But fire—wild and anthropogenic, "good" and unhealthy—is always one step ahead of people. Fire invites us to reimagine the structures and systems we have been living inside and, when necessary, to dismantle them.

In a best-case scenario, the present-day prospect of living with fire is an opportunity for everyone at risk from the effects of climate change—which is to say everyone—to shift our thinking about the elements of the natural world, moving from fear and exploitation toward respect and responsibility. Barry Lopez called this shift an expression of love. Another word for it might be "relationship." Relationships can be challenging and painful. They can contain complex power dynamics. Like a grove of redwood trees blanketed in ash, they can fill you with despair. They can also be the most beautiful part of being alive, in any season.

Solidarity, Not Charity
Mutual Aid in Chico, California
Hiya Swanhuyser

The word "disaster" tells a story. It is night, the stars shine above, and all is well. But then the sky becomes catastrophically absent of celestial light, and trouble appears. Disaster means "no stars."

In the ancient world, people used stars, those "asters," as tools for navigation on land or sea, and to tell time and season. We saw them as spiritual guides as well. "Your stars" would tell your fate, determine your character, call down blessings—unless you were "star-crossed" or "born under a bad sign." When the stars blink out, it means someone, and maybe everyone, is going to need help.

Dana Williams, a Chico State sociology professor, remembers the day the infamous Camp Fire began in Northern California. It was November 8, 2018. "I woke up that morning and opened my window to the south, and it looked really weird in the sky." It would become the deadliest and most destructive fire in California history, killing eighty-five people and practically wiping the town of Paradise, population 26,000, off the map. As only one of the many deadly and destructive fires of that era, it's a frighteningly serviceable model for the future.

Williams and his partner spent the day deciding whether to evacuate from Chico. Should they put the chickens in boxes in the car and drive to the Bay Area? Ultimately they decided to stay, and their first thought that evening was of a neighbor.

"I remember calling an older friend who would kind of regularly drink a little bit at nighttime, and we called him to make sure he was awake. We were afraid that [if] the fire was coming, he might get stuck in his house." Williams recalls the fire story from his office, sitting in front of an impressive bookshelf, one that reaches higher than seems possible. Bookhenge, a place of learning in the ancient style. The neighbor was wide awake and sober, it turned out. And there is no judgement in Williams's voice.

"But the next day was what I *really* remember," Williams says, still amazed six years later. "Because the whole sky was just bronze-colored." The world appeared out of time, he says, like a vintage sepia-toned photograph. In her book *The Last Fire Season*, Manjula Martin describes the same phenomenon from the previous year's fires in Sonoma County. Wildfires don't just take out the stars, she writes. They take the sun too. "It appeared to be neither day nor night. The redwoods were dreamscapes, umbrous black against a glowing red sky. Although they appeared shadowlike themselves, the trees cast none of their own. There wasn't enough sunlight or contrast to make shadows."

Skylessness, or a sky gone wrong, is now known to most Californians.

"We live in the disasters of colonization and capitalism every day, and it's these systemic disasters we spend our time responding to after the embers have gone cold or the waters clear," reads a workshop announcement on the website of Mutual Aid Disaster Relief (MADR). "Earth's natural cycles aren't the problem. The disaster is the way institutions capitalize from and create inequality. It's the power structure that holds a monopoly on aid, but refuses to distribute it to those in most dire need."

MADR, a nationwide network with the goal of supporting communities and helping them heal after catastrophe, has a motto: *Solidarity, not charity*. Aid, the reasoning goes, flows in all directions, not just from the rich to the poor, or from the housed to the unhoused. The network supports community organizing and efforts to teach and learn communication skills, but it also functions as a clearinghouse for practical information. "Composting Toilets, an Introduction," "Chainsaw Safety," "Como Construir un Filtro de Aire de Bajo Costo," and "DIY Emergency Handwashing Station" are featured tutorials on its website.

This hands-on approach was fresh in people's minds in November 2018—Williams included. "[People in Chico] had recently come into contact with the mutual aid disaster relief organization. In fact, there had been a training a couple weeks prior at the Blackbird," Williams says, referring to the treasured local bookstore and café, which did not survive the pandemic. The people who attended that training, and who also came to a huge, unruly meeting immediately after the fires began, had a basic common understanding.

"Many of them, I think, had some kind of general critique of how nonprofits and other kinds of organizations, while sometimes helpful, can be problematic in different ways," Williams says diplomatically. One of those ways is a matter of simple logistics. Neighbors are always the *first* first responders, simply because they're already there, but larger organizations are just big ships. The local newspaper, the *Chico Enterprise-Record* reported over two weeks after the fire started: "Amanda Ree of the American Red Cross in Chico noted that financial donations can be made online, but can only be made to California wildfires in general, not specifically to Camp Fire relief. 'We don't have the organizational nimbleness to do that,' she said Monday."

At that "crammed" initial meeting at Blackbird, Williams and many others formed North Valley Mutual Aid (NVMA).

"We have begun organizing around the core values and principles developed by Mutual Aid Disaster Relief (MADR)," reads an NVMA announcement from November 27. "We have open group meetings which have given rise to several working groups taking action here every day." Their collection of committees, dedicated to the idea that those in most need are best able to decide what they need, first listened to the gathering evacuees.

Then began the child care, cooking, bureaucracy help, sanitation, and more. Crucially, the all-volunteer crew took a mutual aid approach to who they would help. "We are also not making a distinction between 'displaced' and 'homeless' persons. There has been a lot of discrimination against those perceived as long-time homeless with assistance being denied."

Mutual aid seems like a new idea. Most people today would tell you they look to the authorities for help in a crisis, instead of (or more likely in addition to) calmly rolling up their sleeves to pitch in, knowing they can rely on whoever is nearest for aid if they need it. MADR and like-minded groups facilitate what for many is a light-bulb moment: "Wait a minute! We can do a lot on our own." But mutual aid is also a very old idea.

The precise phrase "mutual aid" is sometimes credited to Peter Kropotkin, a nineteenth- and twentieth-century Russian anarchist philosopher and scientist. Taking Thomas Hobbes–inspired Social Darwinists as his main opponents, in 1902 he wrote *Mutual Aid: A Factor in Evolution*. After long-windedly describing cooperative behavior in Siberian animals, Kropotkin gets to the point, humanity-wise: "There always were writers who took a pessimistic view . . . they concluded that mankind is nothing but a loose aggregation of beings, always ready to fight with each other, and only prevented from so doing by the intervention of some authority." Kropotkin disagrees at length with such thinkers and instead lays out voluminous evidence for "the great principle of Mutual Aid which grants the best chances of survival to those who best support each other in the struggle for life."

Yet in concept and in use, mutual aid is much older and not European. In addition to cooperative, nonviolent, and sharing-based Indigenous societies all over the planet, in the United States the term "mutual aid" was in vigorous use by African Americans well before the Civil War. According to Jessica Gordon Nembhard's *Collective Courage*, "Black mutual-aid and beneficial societies spread rapidly in the early 1800s, especially in the North and in urban areas. . . . By 1830 there were more than a hundred mutual-aid societies in

Philadelphia alone. . . . In 1855, 9,762 African Americans were members of 108 Black mutual-aid societies in Philadelphia."

Although many of those societies were restricted to male membership, women cooperated on their own. In the 1790s, Nembhard's research shows, "Black women established day nurseries, orphanages, homes for the aged and infirm, hospitals, cemeteries, [and] night schools. . . . In creating autonomous institutions to solve the problems cause by inadequate health care services, substandard housing, economic deprivation, and segregated schools, Black women served notice that they felt a special responsibility to provide social welfare programs for their communities." They did it themselves—DIT. In the process, they left inspiration for the rest of us.

Raymi Ray calls in from the drivers' seat of a car. They wear a dark knit shirt, the kind often used as a uniform, and they're at the end of a long day of physical work. Still, spiky, bright hair and an ultra-red beanie reflect an energetic individuality. In the back seat of the car, a sandy-colored dog named Shade waits patiently.

Ray, a resident of Chico, had been out of town at a memorial when the Camp Fire began and raced home. Asking around about where to volunteer, they ended up at the center of what would come to be called Wallywood—located in what travel blogger Darcie DeAngelo calls "the modern capitalist commons: the Walmart parking lot."

During the Camp Fire, thousands of evacuees filled the now-infamous lot; many would outstay Walmart's customary informal welcome, even after the mega-retailer hired threatening-looking private security to discourage them. Most had lost their homes and had nowhere else to go. "People coming down from Paradise knew that you could park there overnight, I think," Ray says.

"So I got there and started volunteering; I was either sorting through goods or serving food or just being a supportive ear for people."

"So much of our work was just trying to manage all of this stuff," they say, ticking off questions on tattooed fingers: where to store perishables, how to keep them from going bad, how to make them into food.

The *Enterprise-Record* reported in a November 21 article: "Well-meaning people have been asked to stop sending material donations for Camp Fire victims and to send monetary donations instead."

"Because now we have a pallet full of this one thing," Ray remembers. "And without any communication with any of us, there's just all this stuff thrown into this chaotic situation."

On top of the lack of communication between those in need and those with donations to make, it was telling to see and touch the otherwise hoarded abundance. "The amount of resources that arrived was over the top." It forced reflection and Ray wondered: where exactly was all that help, that aid, on a normal day? If all this could be donated so quickly, why did so many people remain so perpetually desperate?

"For me, I was like, oh my gosh, *any* trauma could make you not be able to stand on your own two feet. Literally any trauma. And here's this really big example of a really dramatic thing that affects a bunch of people, but you can say the same thing about somebody else in the same place, without a house or struggling. Like, we don't need to know the trauma, we don't need to know the struggle that brought them there."

Ray and Williams both report challenges in the process: burnout, confusion, wasted time. But each also, somewhat squeamishly, recalls unexpected joy. "I'm glad that I got connected because it is really—it was a really good period of life," Ray says. "And under such duress. I mean, I wouldn't want that to ever happen, of course. But that was a cool experience."

"There's nothing quite like being in a group, amongst a group of people who are trying to do important things that are good, and do it in the way that you think they should be done," says Williams. "It's very rewarding, and it's very human and very real, and, I mean, it was great." Ray especially seems to feel—understandably—conflicted about the positive side of disaster relief. "Sounds kind of gross to say it like that. But you know what I mean." In truth, they've hit on a confusing but profoundly integral element of mutual aid: for aid to flow in multiple directions, helpers must also be helped. And so, in the NVMA and so many others, they receive joy. Everyone who has helped during a crisis knows what Ray means, although it's likely everyone also feels somewhat weird about their disaster joy.

Ray, it turned out, had not heard of *A Paradise Built in Hell*, a book dedicated to "the extraordinary communities that arise in disaster." Its author, Rebecca Solnit, presents example after example of kindness and solidarity in catastrophes throughout history and all over the world. "As so often happens in disaster, people need to give," she writes. "And giving and receiving meld into a reciprocity that is the emotional equivalent of mutual aid." She quotes Deborah Stone in *The Samaritan's Dilemma*, as well: "Most people don't experience altruism as self-sacrifice. They experience it as a two-way street, as giving and receiving at the same time." And historian Temma Kaplan, a participant in the Freedom Summer protests of 1964, takes it even further, to a utopic place, a sacred place, sprouted from nightmare. "For a short time, during the first few days after 9/11, I felt that 'beloved community' that we talked about in the civil rights movement."

In Chico the newspaper declines to acknowledge NVMA or use the phrase "mutual aid," preferring to simply record nameless "volunteers," and its reporters file familiar-sounding stories. On November 16, long enough for the danger of flames to have given way to humanitarian disaster, in a story headlined "After Being Asked to Leave, Campers Remain at Chico Walmart," journalist Risa Johnson writes:

> Steve Bigalow, a 58-year-old Butte Valley resident, stopped by the donation hub to pick up a few items of clothing. Bigalow has been sleeping in his car outside of the Chico Lowe's with Dana Kenton, his fiancé of seven years, age 57. They don't know if their house is still standing.
>
> Bigalow said he has been blown away by support from the community in terms of donations and other acts of kindness.

For the past few days, a Chico woman—a stranger—has sought out their car and offered them money and home-cooked food.

"I've never seen anything like this in my entire life," he said, getting teary-eyed.

Asked why they weren't staying in an evacuation center, Bigalow said he didn't want to take a space away from somebody else. They have blankets and turn on the heater when the cold gets unbearable, he said. The temperature has been dipping below 40 degrees at night in Chico.

"There's people out there who need it more than us," he said, adding that they didn't have kids to take care of.

Williams notes his appreciation of Solnit's book: "It's so beautiful!" And it is accurate, according to his own experiences. "The thing that happens in crises is that people stop—I mean, I don't want to put this too bluntly, but they stop being assholes. They stop being callous and individualistic and selfish, and their generosity comes out. I think they're a little bit less guarded and particular about what they think should or shouldn't happen."

The "solidarity, not charity" philosophy that animates mutual aid asks what might happen if the false binary between deserving and undeserving melted away permanently.

Because when this happens where you live, what will you do? The climate crisis will bring disaster to your door, whether it already has or not. You yourself are likely to get homeless, to have no place to go, to lose everything you own. That's why it's called a crisis, not just "change" anymore.

"People experiencing homelessness" is a phrase coined to point from now to then, to when you have the experience. On April 15, 2024, a news article about a supportive housing project in San Francisco elicited a sour, telling comment: "I fell [sic] like this will cause further issues for real residents in the area."

As a result of those sour attitudes, people experiencing homelessness are routinely denied disaster aid. The binary between "real residents" and the unhoused is false, but it persists.

Interestingly, during acute disasters, the binary is documented again and again to disappear for a short time, while large numbers of people need help and large numbers of people give help, spontaneously and reciprocally. Authorities are often yet to arrive on scene, and everyday people don't check IDs, they just give and feel lucky to do so. One example comes from Karen Grigsby, writing in a Tennessean newspaper about a photographic retrospective of flooding that had happened in 2010. She found "stories of residents rescued by Nashvillians who dubbed themselves 'The Redneck Armada' and used their personal boats to ferry strangers to safety, oftentimes not even exchanging names with the people whose lives they saved." This kind of help has countless iterations, around the world and throughout history, according to Solnit's book. Even in the same disaster, in fact. Grigsby continues: "These kinds of scenes weren't uncommon in the aftermath of the flood. There was an outpouring of aid, with 29,000 volunteers providing more than 375,000 service hours distributing water and cleaning up homes."

On the other hand is the idea of taking the odd beautiful space of neighbors helping neighbors and making it the new normal. Expanding the moment when people stop being assholes, when kindness feels like safety, until it's like a sky.

It's a project already underway. Solnit points to programs like NERT, San Francisco's Neighborhood Emergency Response Team, which empowers your neighbors to know how to shut off water, gas, and electricity mains, be versed in CPR, and how to search for and carry the injured. Other cities have similar programs, and those that don't can find similar practical information via, of course, Mutual Aid Disaster Relief.

Dana Williams explains that every meeting at Blackbird got smaller, and eventually the loose organization evaporated. This, it turns out, is part of the process of mutual aid, a constant growth, dieback, and regrowth of spontaneous associations. Williams noticed this rhizomatic action: "I think that other organizations picked up some of the mantle a little bit; there's an organization called NorCal Resists... overlapping with the time of the Camp Fire, the Democratic

Socialists of America had mutual aid projects too. And I wouldn't be surprised if there were a few other organizations that either did have similar kinds of efforts, probably smaller-scale, that may have been inspired by ours, or that still are doing some of them too. Sometimes they don't even have names. I think there's sometimes just groups of people who are, like, 'Hey, this is important for us to do.'"

Raymi Ray has a habit of placing a hand flat over their chest. It's a gesture that might mean many things—emphasis, fatigue, a plea. But in Ray's case it appears to signify something like a pledge or hope. It seems unconscious. It's there when they envision the generosity that they saw up close, growing to cover those most in need, all the time. "It was so much abundance. And you're trying to sort it out and like get it into the people." Ray smiles and gives advice to wealth and resource hoarders. "Just give it out, like, let it go." It sounds like it would be a relief.

Farther away and with an inspiring, grand scope, the Sarvodaya Shramadana Movement in Sri Lanka is in action with the goal to "create a no poverty, no affluence, and a conflict-free society." Its many networked village groups include representatives who are specifically children, mothers, elders, and adults from each locality who offer and participate in workshops on micro-agriculture, hate-speech and religious-intolerance prevention, well renovation, and creation of "disaster resilience centres," with the motto "We build the road and the road builds us." Many of Sarvodaya's projects focus on entrepreneurship yet also on "intentional mutual benefit." It's not about getting rich or wielding power.

Stars will continue to shine, of course, but as the climate crisis and systemic disasters progress, they'll also continue to go out. Can we keep the sky clear and spangled? We are headed for disaster—but neighbors, friends, relatives, and strangers always seem to show up and shine some light of their own, one on another.

UndocuFund
Grassroots Solidarity Networks of Mutual Aid for Liberated Futures
Beatrice Camacho

It's nine in the morning, just an hour before the start of the first UndocuFund Encuentro of 2024, only the second of its kind for UndocuFund. These quarterly gatherings are crafted to bring together Sonoma County's UndocuFund recipients in a collective space. As the UndocuFund support team sets up pan dulce, coffee, tea, chairs, and the welcome table, the sound of the door opening signals the arrival of the first family, an hour ahead of schedule. UndocuFund Encuentros have evolved into more than just informational meetings; they've become welcoming spaces for families—intergenerational spaces where they can be seen, heard, engaged, and educated about issues affecting their lives.

Stepping through the doors of the Encuentro venue is a *señora*, accompanied by her husband, sister, and daughter. With a mix of curiosity and hesitation, she asks in Spanish, "Is this where the informational meeting is taking place?"

Through phone calls and text messages, UndocuFund recipients are invited to these Encuentros to connect with others in the community and learn about their rights as undocumented renters, workers, immigrants, and more. They also get insights from community organizers and leaders about ongoing campaigns relevant to their lives, all while sharing a meal and experiencing healing practices like massages and herbal workshops, which have become favorites.

Just as the *señora* asks her question, her daughter, who is about twelve years old, excitedly yells "Rocio!" UndocuFund's coordinator, Rocio, is known by just about every Latinx community member in Sonoma County. So the joyous reaction is nothing new. Rocio warmly embraces the young girl and then the *señora*, as her hesitancy melts away into warmth and familiarity. The *señora* shares that her daughter attended the school where Rocio used to work years ago. With a smile, she adds in Spanish, "Now I know I'm in the right spot. Can we help with setting up?"

All UndocuFund recipients have endured various forms of disaster, spanning wildfires, home fires, floods, power shutoffs, and the relentless impact of COVID-19. While each experience is unique, a common thread of resilience binds them together.

A Personal Experience

"The worst wildfire in California history." This statement has unfortunately become all too common in the past decade. Every year I hope not to hear it. Yet our "worst wildfire in California history" struck the place I've called home my entire life. It was a Sunday night in October 2017. That night, and the days and months following, will forever stay ingrained in my memory—the destruction, the fear, the unknown, the injustices, and the community mobilization and care.

I remember having my bedroom window open sometime between 10:00 p.m. and 11:00 p.m. when suddenly I started smelling what I thought was someone barbecuing. I wondered why someone was barbecuing so late. After looking out my window, with no luck in figuring out where the smell was coming from, I shut my window, closed my blinds, and crawled into bed.

I was awakened sometime around 2:30 a.m. by howling hurricane-level winds that were making my bedroom window rattle. Something didn't feel right. I got out of bed and looked through my blinds out into my neighborhood in Northwest Santa Rosa. The neighborhood looked like it was covered in a red and orange light fog, while the trees swayed from side to side, as if they were going to bend in half. I had never seen the trees in my neighborhood sway back and forth as vigorously as they did that night. Something definitely wasn't right.

    I grabbed my phone and began to Google. "Santa Rosa, California." Nothing. "Santa Rosa, California, wind." Still nothing. Then my phone rang. There was a fire. Where? Nobody knew for sure, but it was nearby. I quickly woke everyone in my household. We all got in one car and drove to a nearby family member's house to figure out what was happening. Little did we know that we were driving toward what we would later learn was the Tubbs Fire. Embers were flying, and neighbors were saying that the fire was coming from the northeast, so we needed to move. We were five minutes away from Coffey Park, one of the areas that would be hit hardest by the Tubbs Fire, with thirteen hundred structures destroyed. As we made our way south, I continued to look for any information that I could find. Where was the fire, and which direction was it moving? Was any of it contained? Where were we supposed to go? How were people getting information? It was either coming from the east or the north, people said. We were hearing different stories, stories of gas stations exploding, buildings burning down. Some of those stories ended up being true, others didn't.

    When we left our home, we each grabbed a change of clothes, assuming that we would only be gone for a short time. As soon as we realized the severity of the fire, my mother broke down in tears. She is a retired domestic worker who speaks only Spanish and came to this country undocumented, crossing the Rio Grande. She was crying because she hadn't brought the money she had been saving, money that represented her hard work and dedication. I remember comforting her, telling her not to worry, because everything was going to be okay, even though I didn't know if it was.

    I was evacuated for a week with family members that I was able to stay with. My home was still standing when I went back to it one week later. It was one of the most difficult and uncertain experiences I've gone through. And I had the privilege of a safe place to go. I spoke English. I had people that I was communicating with to access information, even if there wasn't much information available. What about folks who didn't have somewhere to go? Those who didn't have the time to grab anything and lost everything? What about those who didn't speak English? What was it like for them?

Community Care and Mutual Aid

We know that a wildfire will burn anything and everything in its path; it does not discriminate. But we do have control over how communities respond during and after a wildfire to ensure that *we* do not discriminate. While the Tubbs Fire was burning across Sonoma County, local grassroots and base-building organizations—North Bay Organizing Project, Graton Day Labor Center, and North Bay Jobs with Justice—quickly came together alongside other organizations and community members to discuss what folks were seeing and hearing on the ground.

What kept rising to the top was the impact that the fire was having on undocumented folks in the county—folks who didn't have a safety net, folks who were unable to access information because it wasn't being provided in the language that they spoke. It was because of the longstanding and trusted relationships that these three organizations had cultivated with undocumented folks that they were able to have a pulse on what our front-line communities were experiencing.

Many undocumented folks were evacuating to the coast, toward the water, in hopes that the fire would not reach there. Some were renters who lost their homes. Many were workers who were unable to work—farmworkers who work the vineyards that produce billions of dollars' worth of wine, or domestic workers who clean homes and hotel rooms that are a vital part of Sonoma County's tourism. They were families living out of their cars.

Those who went to one of the local evacuation shelters did not feel welcomed because the individuals staffing the shelters only spoke English. What the folks of those three organizations knew was that it was a choice to not include language justice in life-saving protocols and that it is a choice to exclude undocumented folks from accessing recovery resources.

Knowing that undocumented folks would not have access to recovery resources and efforts, North Bay Organizing Project, Graton Day Labor Center, and North Bay Jobs with Justice created the UndocuFund for Fire Relief, also known as UndocuFund—the first fund in the country developed to provide emergency aid to undocumented residents during times of disaster. Undocumented individuals filled out applications for the fund through one-on-one intakes with trained community volunteers from one of the three founding organizations.

Applications were processed for impact and need, ultimately providing checks of up to $3,000 to families who lost homes. Qualified recipients received more mutual aid assistance. Recipients were able to decide how to best use their funds—housing, repairs, medical costs, food, et cetera. These funds were the only source of support for many folks.

A total of $6 million was raised and disbursed to Sonoma County's undocumented families impacted by the Tubbs Fire. These funds came from all over the country and included small and large donations. Each dollar donated went directly to undocumented families.

UndocuFund was intended to be a one-time fund that would ensure that Sonoma County's 38,500 (at the time) undocumented community members had access to monetary mutual aid assistance during the 2017 Tubbs Fire. Yet, due to additional wildfires, power shutoffs, floods, and the lack of real resources for Sonoma County's undocumented community, UndocuFund has been reactivated in response to each new disaster. During the first year of the COVID-19 pandemic, California's unemployed citizen workers were eligible for up to twenty times as much economic aid as unemployed undocumented workers. When the pandemic hit, UndocuFund once again mobilized to raise funds to get them in the hands of undocumented folks in need.

Following a year of extensive deliberation, the founding organizations, now comprising the steering committee of UndocuFund, came to the pivotal realization that relying solely on direct mutual aid was inadequate. It became apparent that to create real change, organizing and advocacy efforts needed to be part of UndocuFund's work. Therefore, the steering committee made the strategic decision to transition UndocuFund from a reactive model to a year-round initiative and hire a director. I came on board as the first UndocuFund director in January 2022.

For UndocuFund, community care extends beyond the local work. When individuals from Santa Barbara and Ventura Counties reached out to UndocuFund's founding organizations after being affected by the December 2017 Thomas Fire and mudslides, they recognized the necessity of establishing an UndocuFund for their own communities. Sharing UndocuFund's procedures, protocols, trainings, and forms of solidarity became a means of expanding our mutual aid network. In January 2018, the second UndocuFund, 805 UndocuFund, was established.

Little did individuals in Sonoma County know that additional climate disasters and a global pandemic would soon thereafter mobilize grassroots organizations, educational institutions, and communities to create over thirty UndocuFunds across the United States.

Building Bridges and Solidarity

In early 2022, shortly after starting in my role as director of UndocuFund, I had a phone conversation with Maria Melo, then executive director of 805 UndocuFund. She had recently stepped into her new role after 805 UndocuFund became its own 501(c)(3) at the end of 2021. I recall our conversation quickly evolving as we learned about each other's backgrounds. Maria told me about having been born and raised in Colombia, and I shared my parents' story of migration from northern Mexico to Sonoma County and their journey to citizenship (my mother and brother came to this country undocumented). As we learned more about each other and why we do this work, we discovered a common theme: the belief that every individual, regardless of status, deserves what they need when impacted by a disaster.

We began brainstorming. How do we expand? How do we support one another in our work? Having witnessed firsthand the impact of our efforts on the community, not just for those receiving mutual aid but also others beyond, we wondered, "What if we brought together organizations that have created their own UndocuFund? Organizations that have mobilized in response to wildfires, floods, COVID-19? What would that space look like? What would it feel like? What would it bring up for folks?" The more we dreamed, the more we realized the necessity of this space. We needed it.

We decided to convene organizations from across California that shared our values. We began having one-on-one conversations with staff from undocumented and immigrant-serving organizations across California. The response was enthusiastic. Such a space didn't exist—a space where we could work toward meeting the needs of California's immigrant and undocumented communities impacted by disasters. Soon we were planning the California UndocuFund Summit. We chose to host the summit in Santa Rosa, where UndocuFund began.

In September 2022, one month shy of the fifth anniversary of the Tubbs Fire, we brought together twenty-eight California-based organizations that mobilized in support of California's undocumented communities impacted by disasters. California is home to approximately 2 million undocumented individuals and approximately 1.1 million undocumented workers. It's a state where undocumented workers comprise 6 percent of the workforce and account for $3.7 billion in state and local tax revenue. Given the state's vulnerability to wildfires, floods, and earthquakes, and its substantial increase in major disasters since 2017, the summit was a critical gathering.

For many attendees, the summit marked their first in-person convening since the COVID-19 pandemic began. Our objectives were to facilitate opportunities for individuals to meet in person for the first time and build connections, document lessons learned during the pandemic, explore strategic next steps to enhance fund-raising and direct mutual aid, establish a system to advocate for policies that reduce the vulnerability of undocumented individuals and families in disasters, initiate strategic planning efforts, and provide opportunities for staff members who have been on the front line of disaster response to heal from the trauma of that experience. Attendees recognized the value and importance of this space and deemed it worth traveling to.

There, the California UndocuFund Network was born. A second convening took place in June 2023, when we solidified our mission to transform systems, policies, and practices to meet the needs and ensure the safety of California's immigrant and undocumented communities impacted by disasters. We recognize the need to amplify our work beyond mutual aid efforts alone. In conjunction with our mutual aid efforts, we aim to build a strong and sustainable network for the benefit of undocumented individuals facing disaster.

What has become evident over the past seven years is that we cannot thrive in isolation. Bold collective action is essential. Since 2017 we have witnessed the power of community coming together in the face of devastating wildfires, floods, power shutoffs, and a global pandemic. Through these trials, we have unified, demonstrating a revolutionary spirit, raising over $17 million to directly benefit our undocumented community in Sonoma County. This remarkable achievement underscores the profound impact we can achieve when we unite in solidarity, reaffirming the resilience and strength of our shared humanity.

The Community of Evacuation
Margaret Elysia Garcia

Sheriff Todd Johns made the call, thirteen days before the Dixie Fire ate our town of Greenville, for the bulk of Indian Valley to evacuate. In our little corner of Plumas County, rural California, this meant the towns of Greenville, Taylorsville, Crescent Mills, and the smaller hamlets of Indian Falls and Canyon Dam. Later that was extended to Lake Almanor, Genesee Valley, and the outskirts of Chester. For a while East Quincy and the area around Mount Hough and Keddie Ridge were under evacuation. Meadow Valley. Butterfly Valley. It was too many of our towns at once. Two-thirds of our mostly forested county burned in the nearly three-month fire. How, where, and when to evacuate became a huge headache for county residents, many of whom did not have the finances to do so.

If you live in a national forest in California, odds are pretty high that at some point or another you've been ordered to evacuate. This wasn't our first rodeo. But no entity had ever asked us to evacuate for this long. One to three days, sure, but two weeks? By the end of it, Indian Valley residents were asked to evacuate for six weeks. But all over Plumas County people were forced to evacuate.

In Indian Valley, for the first twelve days, many of our residents did indeed evacuate, but a significant number stayed behind. Some residents had livestock to look after and often no solid indication of where they could take their animals that wouldn't also need to be evacuated soon. With so many towns evacuating at once, some didn't want to stay in evacuation shelters where the lights would be on all day and night and the likelihood of catching COVID was high. Some people had family or friends to stay with outside of the county and took advantage of that. But for people whose entire families were in the valley, those who never leave their small rural communities, the prospect of evacuation was daunting—and expensive and feeling impossible.

In Plumas County every available hotel, motel, and Airbnb room not in an evacuation area was booked solid for months. People had to push farther and farther out: staying in Reno, Nevada, and in Chico and beyond, maxing out credit cards. It's an expense that low- to moderate-income communities like those comprising most of Indian Valley could not recoup easily. Most could manage one day, two days, maybe three—but twelve? That couldn't be done.

Those who stayed behind did neighborly things like watering and feeding cats that refused to be found when owners had called for them when evacuating. Some had to leave their livestock behind with nowhere to take them. Places to take animals were filling up. Neighbors were temporarily putting their animals in shelters a few hours away.

Some residents who stayed behind tried to water roofs and areas around houses near the foothills. The market stayed open to sell whatever was left to sell without new deliveries. Some residents began camping around the creek if they were too close to the fire's edge. There became bigger camps of people staying on a property behind the high school, a concrete building less likely to ever go up in flames (it didn't, though the newly refurbished gym saw some damage). On North Valley Road on the backside of the valley, a property owner let the guys he hung out with all park their cars out on his property, and they did communal picnic barbecues for as long as they had provisions. This part of rural California has a good strong streak of independence coupled with a "live and let live" attitude and a dash of community spirit; it served us well during evacuation.

Everyone breathed a sigh of relief on day thirteen, thinking the fire would be contained, when Sheriff Johns called off evacuation and let people back into their homes in Greenville. People started to unpack their cars. They let their cats out of their carriers and let the horses back in the fields.

Less than twenty-four hours later, phones started bleating again. Mandatory evacuations were back on. It wasn't looking good in the mountains above Greenville's valley floor, but surely the fire would be stopped at Round Valley Reservoir above Greenville. But the fire was jumping dozer lines. Embers were flying in the high winds across the lake, so the water made no difference. The hillsides were steep and thick with trees and drought-dried underbrush. And the hot, dry wind had picked up. It was time to go.

Much to the dismay of both law enforcement and firefighters, many people stayed too long. It was nearly impossible to get the cats back in the carriers. Evacuating a second time around meant people didn't take everything they took before. People were tired and worn down. They forgot the wedding photos in the box by the door.

Many others moved back into the evacuation zone to help family members and elderly people with their properties. Residents in the Dixie Fire zone knew the fire was huge, knew there were multiple fires going on in the state at once—which meant no one fire in the state was being fought with a full force of firefighters from every agency. Residents knew that the task of fending off the fire from housing structures was going to be gargantuan. In the end, some local volunteer firefighters would wind up watching their own houses burn.

The question continued in everyone's mind in the days leading up to the fire, and in those days after the Dixie Fire brought Plumas County to its knees: do we evacuate? The answer in those days—as it is now—is yes and no. Some evacuated, and some didn't.

Then the fire pushed down the hillside from Round Valley Reservoir, and those remaining could see the fire beginning to take out structures, began hearing the propane tanks blow up like bombs, and headed the hell out of town southeast toward Quincy. One resident cried watching the fire eat up the town from the rearview mirror of her partner's truck. They had stayed till the end, watering and preparing for a firefight, and in the end it had meant nothing. She joined her daughter and her family who had left right after that second evacuation call. The family business—an Ace hardware store in the family for generation upon generation—gone.

Even then there were people who didn't leave Greenville. They couldn't afford to, or they didn't trust the agencies in the emergency relief shelters—or both. There are people who listen to conspiracy theories up here and were sure the fire was a liberal United Nations plot to push people out of their rural homes and into San Francisco. Residents camped out on the valley floor, the only area that wasn't burning, along the trickling creek.

The Town That Said No

One Indian Valley community stands out in its defiance of evacuation: Taylorsville. The tiny town has roughly 150 people, comprising hippie artists, conservative ranchers, and a smattering of people in between, and is known for its independent spirit. The Taylorsville Tavern sports a giant mural of the State of Jefferson flag on one of its walls. It's been no stranger to fires in the past and bears a considerable scar on the hill right above the downtown from several years ago when many thought the center of town would go up. Not everyone remained, but a sizeable number of people young and old alike stayed, long after PG&E cut power. The population stayed after their generators ran out of fuel, and those who stayed did so with a communal spirit. The woman who once owned a pie shop made sandwiches for whoever was left in town; she fed the volunteer firefighters and anyone fighting the fires. The non-evacuation grew increasingly troublesome. Taylorsville residents watched the fire clouds over Greenville where it was burning seven miles away. They knew what could happen. They watered their roofs and decks. They kept their gardens alive. They cut any remaining dead standing trees. They cleared brush if they had any left. They made their own dozer lines. Ranchers with both water rights and water trucks wet down the town from nearby Wolf Creek. They stuck it out.

This was no easy feat after the fire took Greenville. The National Guard had been called in and blockaded the highway that led into and out of the valley. No one could come in to deliver more supplies; people could always leave, but they could not come back. Most people in Taylorsville waited it out. Some snuck in supplies like bandits on dangerous dirt roads.

Much of the defiance seemed to be personified by one man in Taylorsville: Dan Kearns. He began a twice daily livestream on his Facebook page disseminating information to whomever followed him.

"This is a disclaimer," said Dan Kearns at the end of one of his livestream updates. "I'm just a guy. I'm just a volunteer." But he emerged as the go-to person on the ground for information on the Dixie and Fly fires as they headed closer and closer to Taylorsville. He interpreted the daily information releases from the county, the US Forest Service, and other agencies. His increasingly disheveled appearance and beard growing longer by the day seemed to embody what the rest of us were feeling. We'd grown feral. Taking matters into our own hands. His weariness was our weariness. His was the look of the frontiersman. The non-evacuee.

Finding out the latest of what was happening with the Dixie Fire could be more than unsettling and not always easy for the layperson to grasp. Our imaginations worked overtime, and for some of us our worst fears had already been realized. It was very easy to panic.

Kearns became the voice of reason, assurance, and calm in the face of adversity, even if his politics sometimes struck people as bizarre.

"Clear, concise, and easily understood information from just a guy," said Jeanna Van Brocklin on Kearns's Facebook page.

"So awesome to see the community coming together to beat this thing," said Christian Gaston Palmaz, a ranch owner and Napa Valley winery owner who had already lost vineyard property in Napa. There were many, many other thanks on Kearns's page too.

Not only did Kearns give fire updates and explain what was happening at the moment with maps and in terms that everyone could understand, he also came to his livestreams with fire-related information and accolades the public needed to know about too. For example, letting the public know that Alicia Dalton and Wayne Kelly were going around feeding and watering evacuees' pets that may have been left behind. Or that Pappenhausen's Pit Stop and Mary's German Grill were cooking up food for free in downtown Greenville (before it burned) for both firefighters and those who stayed behind. Community members already gone sent electronic funds for others to buy food at the still-open Evergreen market.

He gave shout-outs to those in the community who helped keep the valley afloat as long as they could, tirelessly (and often unsung)—like the Indian Valley Community Services District's main employee, who tried to keep water services up and running. Kearns reminded residents to boil water before using.

The information Kearns shared was all readily available online in social media and official channels, of course, but there was something about both his delivery and his priorities and the calmness of his voice from non-evacuated Taylorsville that had the community appreciating that one of our own was there for us. Toward the end of August, when people still weren't allowed back and provisions were running low, you could tell that Taylorsville residents might be on the brink of madness from the livestreams and the TV crews allowed in to do stories on the defiant town that didn't burn.

Adult children returned to Indian Valley from all over Northern California to help out elderly family members and their properties. There were photos on social media of these returning sons and daughters borrowing bulldozers and cutting their own lines above their parents' and grandparents' properties. They saturating the wood decks and set their own back burns to alleviate fire risk to specific properties, something official firefighters would not have the time or resources to do. It's hard to argue that those in Taylorsville should not have stayed put. None of their properties burned down. They saved their town.

In the spring of 2022, when we all had a clearer picture of what was burned and what was coming back, you could see how far down Mount Hough the fire burned and the huge green swath around Taylorsville that owed its life to those volunteers who didn't leave town. This defiance saved some houses in nearby North Arm and some places in Genesee Valley. For many of us who watched property owners saving their own houses and land and their tiny hamlet of a town, it became hard to argue that evacuation is always the best response to fire disaster.

Evacuation into Chaos

In the aftermath of the Dixie Fire incinerating Greenville, Canyon Dam, and Indian Falls along with the off-the-grid community of Warner Valley, the evacuation was not lifted quickly. News vans descended on the wreckage of the town with press passes to flash at the National Guard, but residents who had left were not allowed back, and those residents became increasingly frustrated. Why were coastal news organizations standing on the bones of the town when we ourselves were not allowed back? It brought an additional sense of hopelessness and loss of control. The governor, Gavin Newsom, visited and declared a disaster, someone from the feds finally declared a federal emergency nearly three weeks into our disaster, when finally enough monetary damage had been done, and the ball began rolling into the chaos of recovery. Still we officially couldn't come back.

For those of us who still had houses, we had no power yet. For those of us not on our own wells and septic, there was no water. The sewer plant had blown up too. The school year began two weeks late for some, a month late for others, in year two of COVID. The evacuation camps at the fairgrounds and campgrounds were full, and people were catching COVID and getting sick from smoke inhalation.

We were finally on the map now that we were burned out of the map. Our downtown had the highest (worst) air quality index (AQI) levels on the entire planet for a week. We clocked in with an AQI of 851 one day.

Evacuation weighs down the psyche. No one can rest. No one can really make plans. Unsympathetic employers had no sympathy for workers burning through vacation time and sick days. Some were evacuated a total of eight weeks in the Dixie Fire, and at the end of the eight weeks many didn't have a place to get back to. Eight weeks is a long purgatory.

We slept in our cars. We took housesitting gigs. We crashed endlessly on friends' couches and couldn't give them an end date. We camped outside even though it wasn't safe to do so. We visited our animals still incarcerated in animal shelters and had no place to take them. We took relatives in Southern California up on that offer to visit them. We grazed meals—random bits of whatever was around, fast food. Kids on evacuation started begging for vegetables and never wanted to see a granola bar again. We hoped for some sort of salvation that felt like it would never come. Kids started school in tents, with internet-based curriculum but no internet access.

No one had made a plan to live like this for so long.

Eight weeks of evacuation—a time when none of us could plan or budget out an uncertain future, any future. We stayed glued to information on Twitter. And even though to not evacuate meant the very real possibility of running out of power and water, it seemed in these moments of uncertainty and unrest that camping out in our smoke-damaged homes was a better deal than whatever this was. To live like this is exhausting. To be checking back online to see if our house was one with an X on a Google Earth map, meaning gone, or a check mark, meaning saved, was its own horrendous punishment during evacuation.

My mother's house had a check mark.

My best friend's house had an X.

My office was leveled to ashes in the first videos posted by fire chasers that I saw on YouTube that first night. When you are in a national disaster, no one checks with you before they post your burning life on YouTube.

My press pass allowed me to go into Greenville, but I wasn't allowed to stay. I took photos and sent them to neighbors when I got back to cell service outside of the burn zone—we didn't have a strong signal in Indian Valley for at least a couple of years. This was my contribution to my community: sneaking in on a press pass for a story I wasn't writing to feed neighbors' cats and take photos of property for anxious elderly people.

These may have been the longest eight weeks of my life. Many of my neighbors and I still talk about it. We talk about the swiftness of earthquakes and hurricanes and tornadoes and flash floods. We talk about how the disaster of fire begins in a spark and is followed by a prayer that the wind blows in a different direction or stops blowing. And then it builds and builds, and even after the fire takes your town you still wait and wait for the world to stop spinning out of control. The anticipation of disaster is longer than the disaster itself. The waiting afterward is just as long.

Evacuation as Community Event

The call came to evacuate for the Dixie Fire. Greenville, Indian Falls, Feather River Canyon communities, East Quincy. Meadow Valley. Feather River College.

There's nothing like a wildfire out of control to make you realize how close mountain communities are to each other even when they seem so far apart. We are just on the other side of a mountain from each other. And that year that mountain was on fire.

When my mother evacuated with her two cats, she went first to a friend's house in Lake Almanor. Her friends made dinner for her, and they all sat around and ate and talked, and then everyone's phones blared an evacuation warning yet again, and she was back on the road to a different friend's house along with the rest of her dinner companions. This time they headed to Susanville in Lassen County and then Sierraville. She couldn't go back the way she'd come. Instead she went the long way around the mountains, nearly to Reno, before doubling back to Plumas County. Her friends loaded up their RV and headed out to a campground near the northwestern part of Susanville.

That took half a day. When she finally arrived in Sierraville, she wasn't the only evacuee. Our friend Linda evacuated from East Quincy to Sierraville. In those crazy two weeks before the Dixie Fire, there were a evacuations of most of our communities at one point or another. We ran out of places to run to with so many of our communities evacuated at once. It was too big for our usual Plan B plans. Over the years, Quincy has been threatened, and people there came to Indian Valley. Maybe we had friends in Chester or Portola—our Plumas County towns are all about a twenty-five-minute to forty-five-minute drive apart. But now evacuations were all over the county. Where to go?

Those of us with family and friends far away from burning mountainsides left the area completely and waited out our time miles from home, glued to our laptops and phones. We spent our time reading Chico-based Zeke Lunder's website www.thelookout.org, where he posts maps and explains fire behavior, likely fire behavior, and never-before-seen fire behavior to both the initiated and the uninitiated. Our Mercury or Hermes, messenger god, always translating for the layperson what we need to know and what we were not getting enough of on the news. The news was aimed toward people far away watching our disaster as morbid entertainment. As warning. If you live in the dry forested mountains of California, you will burn.

Our fires only make the news if the hazardous AQI and orange apocalyptic sky make it to San Francisco or Sacramento. We are news only when the forces of our destruction become a hindrance to the daily lives of city dwellers. In September and October of 2022, the *Los Angeles Times* did a four-part series of opinion pieces called "Rebuild Reburn" in which the two columnists, who do not live rurally, advocated for not rebuilding rural California when it burns because it costs the real California taxpayers too much, and we rural folks are just a bunch of hicks who don't vote correctly anyway. I wrote to them to try and reason with them. Where is this permanent evacuation

Shangri-La they would have us live in? Rents in Plumas County still run under $2,000 a month—nearly unheard of in California. Before the Dixie Fire, rents in Greenville were under $1,000. This is part of what makes evacuation nearly impossible for anyone without family and friends outside of the Sierras. Where do low-income people go? Where can we relocate in California? The coastal smugness is something we are accustomed to, but the indifference evident in authorities not presenting a plan of action in their stories hit deep. Everyone has an opinion about fire; no one has an answer. What's more, the Sierras are still home to Native California tribes. Where would the columnists have them go? Are we talking just evacuation? Permanent relocation? Do we not know how American history went down?

My mother, a recent widow who has owned her own homes since the early 1990s and had no idea how much going rents were in California, let alone how much a motel would cost for evacuation, told me before the Dixie Fire that she didn't want to live in the mountains by herself anymore. The winters were hard on her, and the summer fire season was increasingly uneasy. Where could she go and pay $800 a month for a small unit on one floor in California? I searched on Zillow. That will get you a studio apartment in Turlock with nothing included. Motels and hotels begin at $100 a night easy up here. She stayed the remainder of evacuation in Sierraville in an old dilapidated Victorian house full of women who took turns cooking, laundering, and doing other duties while they enjoyed conversations together and read books.

It's in this context that property owner Leslie Wehrman opened up her heart to offer thirty-seven people (along with nine dogs, five cats, eight goats, fourteen chickens, and one goldfish) refuge from the fire on her five-acre parcel between the towns of Portola and Graeagle (eleven miles apart), safe space during evacuation.

"Emergency cohabitation with virtual strangers experiencing deep stress and the trauma of loss over a seven-week period was hands down one of the most influential time periods in my life," wrote Wehrman in a story for the *Feather River Co-op Newsletter*. Each evacuee took on a different job in the makeshift commune—from tending the barn of evacuee animals to harvesting and tending the garden, and doing laundry.

Michelle Fulton and her wife Lovely Hatzell of FulHat Farm in Meadow Valley were two of the fire refugees Wehrman took in during what Fulton called "an opportunity for community fellowship." Together with their goats and cat and dogs they took refuge at Wehrman's, though their chickens stayed behind on the farm and were tended to by a neighbor who did not evacuate. Fulton became the barn boss, while Hatzell became the camp chef. Each evacuee took up the role best suited to them. The feeling, the atmosphere the makeshift commune created—all things and personalities taken into account—was calm, slowed down and focused on meeting basic needs and functions. Evacuees didn't try to retain their usual everyday lives. They tried instead to maintain a safe haven for all. According to Fulton, one person eventually left and another person who was a bit high-strung in the first place "freaked out," but by and large they created seven weeks of harmonious living.

"We had fun," said Fulton. "We shared meals. We played games. We had a good time in the middle of all of this." She likened it to a camping trip—if you could get your mind and anxiety into that space instead of one of looming dread and waiting for more bad news to strike.

Perhaps that's what the model for evacuation should be. Like packing go-bags or knowing where the exits are beforehand, what if we had a plan for evacuation that encompassed who we wanted to evacuate with? What if we assessed beforehand what we would have to offer a group in this situation? Seeing evacuation as an opportunity for community might be what saves us all or at least lessens the anxiety and stress of feeling like one is all alone against the fire, every fire.

After Evacuation

As evacuations were beginning to come to a close and residents of evacuated rural mountain hamlets were slowly allowed to return, I got an invitation to join a group on Facebook. It was a group of neighbors around my mother's house. I joined immediately. Nowadays, though my mother and other retiree neighbors have left the state for cheaper and less fiery pastures, the bulk of us remain.

We now act like neighbors should have acted in the first place. From my office at the local newspaper, I begged the sheriff to let us back when the neighboring area was back, pointing out that we were on the same power grid, not on Greenville's grid. We ganged up on Waste Management together when it kept charging us but didn't pick up our trash for weeks on end. We shamed the postal service into restarting service in our area. We let each other know when we saw a bear or mountain lion on our game cameras. We watched each other's property since post-fire break-ins began when the sheriff's department was stretched too thin to do proper patrols. We complained en masse to Frontier Communications and PG&E for their lackluster delivery of services—as monopolies, they are not prone to put a high priority on customer service. Post-evacuation, we've learned a good deal about ourselves. Our humanity came back to us. Our strength was forged in fire.

We are slowly recovering financially. There's no relief coming for those who had to max out their credit cards. For my part, I wish the toll of evacuation was taken more seriously by those around us. Could school administrators and teachers be more empathetic to students whose houses burn, who are not living anywhere stable for the first few weeks of the school year in evacuation? Could landlords, creditors, and employers be more empathetic? Can we all acknowledge that evacuation is a strange planet that evacuees visit physically but not mentally? That it might exist in its own time and space altogether. How do you keep living your normal existence when things are anything but normal?

The answer to the last question may be a single word: "community."

What Wildfires Do to Our Minds
Dani Burlison

In late spring 2018, I hiked Sugarloaf Ridge State Park in Sonoma County with therapist, ecopsychologist, and California naturalist Mary Good. I remember a mist drifting down and that we had the park mostly to ourselves. In October 2017, 80 percent of Sugarloaf's thirty-nine hundred acres of oak woodlands had been scorched by the firestorms in California's North Bay. But on that day with Good, most of what stretched out before us was green and vibrant, brushed with the last signs of a wildflower superbloom that had erupted from the ash earlier that spring.

A dozen miles west in Santa Rosa, contractors were rebuilding some of the more than five thousand homes destroyed there. The last of 2.2 million tons of fire debris had been hauled away from the 383 square miles of charred land in the region. And therapists like Good were seeing fire survivors pro bono, helping them navigate the aftermath of the disaster.

"It was an absolute trauma for everybody involved. The fire is over, but the grief may last a long time," Good said. "We live in a time where these natural disasters are going to be happening more and more. How do you develop resilience? What do you do to feel like you can be safe in the world again?"

Developing that resilience is crucial. According to climate research from NASA, we can expect more droughts, stronger and more intense hurricanes, and big changes in precipitation patterns.

As climate change–related disasters become more common, there is a critical need to address the mental health of survivors after a catastrophe. Santa Rosa residents—and the greater Sonoma County community—rushed in to offer support services through pop-up holistic clinics, mental health education, and free counseling services. It's a response that may help other communities cope with future disasters.

The magnitude and chaos of the North Bay fires left local government and nonprofit organizations overwhelmed—the fires plowed through several neighborhoods overnight, sending more than four thousand people to forty-three shelters at the peak of the fires. The Red Cross and local organizations offered psychological first aid, an emergency response tactic defined by the World Health Organization as "humane, supportive, and practical help to fellow human beings suffering serious crisis events."

National programs can help address people's mental health needs during disasters like these fires and in the immediate aftermath. The Disaster Distress Helpline—a confidential, national 24/7 call and text service—is one such program.

Christian Burgess, the helpline's director, says that most calls during disasters are from people feeling overwhelmed and anxious, seeking information about the event.

"During the long-term recovery . . . we start to see deeper mental health concerns from callers and texters, such as persistent anxiety, depression, and substance abuse, which can be related to traumatic exposure during the event; loss of loved ones, including pets; and financial strain," Burgess says.

Grassroots Mental Health Support

Other organizations in Sonoma County took a more grassroots approach to offer support. At the time of the fires, Tré Vasquez was a youth organizer at the North Bay Organizing Project, a Santa Rosa–based nonprofit that organizes working-class and minority communities to build political power.

When the fires erupted, Vasquez and his team mobilized quickly, collaborating with local churches, herbalists, acupuncturists, ancestral healers, counselors, and community volunteers to launch community healing events called Sanación del Pueblo ("The People's Healing") to support those impacted by the fires, especially the region's large immigrant population.

The first event was hosted within days. In the following weeks and months, Sanación del Pueblo provided physical and emotional support, referrals, and meals, and donated respirators to nearly six hundred people. The events were hosted at a community garden, a local Unitarian church, and a branch of the Sonoma County Library in a largely working-class and Latinx neighborhood of Santa Rosa. North Bay Organizing Project (NBOP) continued offering events on a quarterly basis for at least a year, and the events included people of all ages sharing meals or chatting while waiting for their turn at massage tables, counseling sessions, or *limpias*—traditional Mexican spiritual healings.

According to Vasquez, Sanación del Pueblo centered people who have been historically underserved by medical providers, including undocumented immigrants, women and trans people, those with existing mental health concerns, and others at risk of being left out of emergency response services.

Other local organizations and individuals stepped up to provide emergency relief after the fires too. In central Santa Rosa, the Lomi Psychotherapy Clinic—a sliding-scale outpatient mental health clinic—opened its doors to fire survivors immediately, advertising drop-in services over local radio to draw people in.

Thomas Pope, Lomi's cofounder and clinical director, says the clinic initially saw about fifty new clients in its fire survivor program. The program offers free and reduced-fee counseling services and was partially funded by the North Bay Fire Fund, which raised over $32 million in four months after the fires. Pope and his staff of roughly thirty therapists hoped to provide services to survivors for as long as they need them.

"What we know is that three months to a year after a disaster is when the most need happens; that's why we want to keep this going," Pope says. "I think it's going to be quite a while until this community finds its way out of this initial stage of shock."

Pope's advice to other communities responding to large-scale disasters echoes NBOP's actions: create safe places for people to go where they will have connections with others and positive activities to focus on. He says that finding a balance between discussing what happened and engaging in activities that bring pleasure and nourishment is key.

"Looking at disasters and the wide range of traumatic response, it's really good for our communities to know that there is a huge range of response," Pope says. "And it's important to attend to all of it."

Pope says that providing services as soon as possible should also be prioritized. Immediately after a disaster, people need help navigating resources, calming themselves, and problem-solving—all key aspects of psychological first aid. Trainings in psychological first aid are available online through organizations like the *Substance Abuse and Mental Health Services Administration* (SAMHSA), and anyone can attend. For survivors, having trauma validated and finding a supportive environment quickly can be critical for long-term well-being.

"And we really need to learn in recovery, to be able to shift attention away from difficult things to what's working well: love, connection, beauty, and joy," Pope says. "I don't want to sound callous at all, because in the middle of trauma, we can't always do that. But in the short range we also need to learn how to get out of the well of despair and find goodness also. And that's what we saw in this community: there's an amazing amount of goodwill and care and love and goodness. That's part of recovery: being able to allow that support and to internalize the care that is here."

Throughout Sonoma County, other support networks have surfaced, including free trauma-informed yoga classes, support groups through hospice organizations, brown-bag lunch discussions, presentations on how to recognize and support loved ones with post-traumatic stress disorder, and holistic health care providers offering free services. But as the land regenerates and homes are rebuilt, the traumatic memories and uncertainty of being unhoused remain painful realities for many.

David Leal, a US Navy veteran, used many of these services immediately after he and his wife lost their home of ten years in the Coffey Park neighborhood of Santa Rosa. He attended a free yoga class for fire survivors three days after the fires started.

"The instructor was very compassionate and offered her support at no cost," Leal says. "It was my first lesson in receiving help."

Leal also attended the first Sanación del Pueblo event, where he received a free massage and herbal supplements that he continued using. He also continued a regular yoga practice and received free and low-cost acupuncture and herbal supplements that helped him immensely with the service-related PTSD that was reignited after losing his home.

"The fire triggered a lot of old stuff that I had experienced all the way back to childhood. The greatest challenge has been loss of sleep due to dreams and nightmares of so many different painful episodes from my past," Leal says. "But between yoga practice and chats with my navy psych friend—and the herbs—I've been able to recover from the sleepless nights."

He says that the early support helped him to be calm, especially as he dealt with the stress and red tape of rebuilding his home.

Mutual aid efforts like those in Sonoma County help create a strong sense of community resilience. And when fire survivors step in to add to the support efforts, it can help them further along in their healing journeys.

Helping Others, Helping Self

The physical and psychological benefits of helping others continue to be studied, but research published by *Psychosomatic Medicine: Journal of Behavioral Medicine* suggests that helping others contributes to brain health and emotional well-being. The study also revealed that providing targeted support to specific people activates the region of the brain associated with parental care and is more beneficial to emotional well-being than donating money to an organization.

A similar study published in the *International Journal of Behavioral Medicine* in 2005 reported that "a strong correlation exists between the well-being, happiness, health, and longevity of people who are emotionally and behaviorally compassionate." The study had a caveat, however: offering support is only beneficial if the helper is not overwhelmed by the tasks.

Helping others with similar experiences is not a cure-all for trauma. According to the 2005 study, people who are physically or mentally overwhelmed or feel taxed by the needs of others can experience significant strain that can contribute to negative health consequences.

"What we do know very clearly: previous trauma affects a person's capacity for resilience," says marriage and family therapist Linda Graham, author of *Resilience: Powerful Practices for Bouncing Back from Disappointment, Difficulty, and Even Disaster*. "If there's been previous trauma, it affects the capacity to deal with a catastrophe, to bounce back from a catastrophe. . . . But even if people have had traumas, if they've had the strength of resilience to navigate that, then they're less likely to go into trauma the next time something difficult happens."

According to the American Psychological Association, resilience is "adapting well in the face of adversity, trauma, tragedy, threats, or significant sources of stress" and bouncing back from challenging experiences. Being resilient does not mean that a person won't experience negative emotions in the face of stress, however—it means managing those feelings in order to carry on with life in a healthy way.

Pennisue Hignell is the director of the Northern California Trauma Recovery Network, a group of eighty-five therapists who offered pro bono sessions to address early trauma from Northern

California fires like the November 2018 Camp Fire in Paradise. She says that in addition to the eye movement desensitization and reprocessing treatments that she and others offered, providing social support can indeed benefit the brains of survivors and support them in the healing process.

Hignell says social support helps develop "the adaptive brain," a region of the brain that psychologists attribute with resilience and combatting emotional distress.

"The adaptive part of the brain is developed through spirituality, friendships, relationships, someone who loves you, meditation, yoga," she says. "Giving service to others develops things that put endorphins in the brain. So that person feels better, I would imagine, in serving [others]."

Hignell also points out that simply connecting with community can have a positive impact on survivors. "I think that it would help them a lot because I think the hardest thing for survivors is they lost their community," she says. "That's far more significant than losing a house."

Danilla Sands lost her home in a single residential fire when she was eleven and later watched as family members' and loved ones' homes perished in more recent fires around Northern California. She also lost a close friend in the 2017 Redwood Valley fire.

Since the 2015 Valley Fires in Lake County, she has been a volunteer coordinator for fire relief services, including managing a networking page on Facebook and establishing two donation sites, which she voluntarily coordinates year-round. Sands says that of the countless volunteers she interacts with, fire survivors are the first to step forward to help. She initially felt conflicted about accepting help from people who had lost so much but has come to see their efforts as part of their recovery process.

"The most beautiful thing I have witnessed is a volunteer or previous fire survivor hugging a recent fire survivor and reassuring them things will be okay," she says. "I know these families have the strength to get back on their feet. Sometimes they just need someone to lift them up until they're ready."

It's no surprise that low-income and other marginalized communities are disproportionately impacted by climate disasters like the frequent firestorms we experience here in Northern California. Many referred to the 2017 Tubbs Fire in Sonoma County as a great equalizer, as it burned through a variety of neighborhoods in Santa Rosa, from the wealthier, hilltop community of Fountain Grove down through the Journey's End mobile home park, and across Highway 101 to the middle- and working-class neighborhood of Coffey Park. While people in each neighborhood suffered total loss of property, we know that the ability to recover was much more difficult for those with the lowest incomes.

The United Nations Office for Disaster Risk Reduction says, "Disaster risks in rural areas may be particularly invisible, given the low density of produced capital and declining population. Poor rural livelihoods are highly exposed and vulnerable to weather-related hazards and have a low resilience to loss because they have little or no surplus capacity to absorb crop or livestock income losses and to recover. Even a small loss might feed back into further poverty and future vulnerability." All of this can contribute negatively to mental health.

An additional negative mental-health impact—speaking as someone from a poor, rural family from Tehama County, growing up with deeply loved connections to Butte, Lassen, and Shasta counties—is being treated as if the people living in these rural areas are disposable. The *Los Angeles Times* ran a series of stories in September 2022, criticizing rebuilding efforts in Greenville as it began the long road to community recovery after the Dixie Fire burned down the town.

Yes, some question the advisability of rebuilding towns in fire-prone regions. These are regions that, according to the US Forest Service, saw an increase from 30.8 to 43.4 million homes (a 41 percent rise) between 1990 and 2010. The area of Northern California where Paradise is located is one such fire-prone region. As climate change continues bringing higher temperatures and lower precipitation throughout California, fire seasons are projected to get worse throughout the state. But Paradise and Butte County in general are part of a largely working-class region. According to a 2016 Butte County Health Assessment Report, the county's median annual income was roughly $43,000, and nearly 60% percent of children were eligible for free or reduced-fee school lunch programs before the fire. For many, moving into more expensive areas of California, where there continues to be an extreme shortage of affordable housing, is not a feasible option.

One person who hoped to rebuild his home—and whom I met at a permaculture restoration camp in Paradise in 2018—is a man known as Pyramid Michael in the Paradise community. In his seventies, Michael is a veteran and construction worker turned massage therapist, and he spent ten years designing and building an energy-efficient, passive-solar-powered home in Paradise. Before the Camp Fire, he did a "permablitz"—a comprehensive permaculture project—on his property that included planting a garden and small food forest and installing a rain catchment system.

"Then the fire came through and wiped it all out," he says. "But I've been homeless many times in my life. I know what it's like to be without nothing or starting over again. But I'm still healthy. I have strength, and I have intelligence. And I have a vision. And I know how to work with those."

I'd argue that what people from communities like Paradise and Greenville need is community support, mutual aid, mental health services, and encouragement. Not criticism for their lifestyles, living conditions, or choices to remain and rebuild in their communities.

"We have a moment right now that's really calling upon us to figure out how we're going to return to living in a good and balanced way," Vasquez said in 2018. "We can create spaces in which the way that we care for each other is a glimpse into the world as it should be. Or as we hope for it to be, as we mean for it to be."

Back to the day at Sugarloaf Ridge, Good said that community training and planning before disaster strikes is a must as communities look toward adapting to the new normal of climate catastrophes. She says that connecting with nature, even after a disaster, is critical, recounting stories of fire survivors regaining hope when the scorched land showed signs of regrowth. Yet she acknowledges that survivors face long roads to recovery.

"Putting an entire life back together—it just stops people in their tracks," Good says. "Where do you even begin? How do you pick a point and start?"

The light rain let up at the park, and Good was excited about showing me a large bay tree that was badly damaged by the Nuns Fire. A hole had been burned through its trunk, but there was new growth sprouting around its blackened base, and leaves springing out from its branches.

"It's such an amazing example of how you can be burned through to your core, both literally and metaphorically [and survive]," she said. "Even after being burned through to the core, [the tree] still leafed out this spring. It's a great example of individual and community regeneration."

Heaven over Fire
Finding Community in a Burn Scar
Jane Braxton Little

The night Greenville burned we threw the I Ching. Jon and I huddled together on Jane Chang's vintage blue-and-red-plaid love seat. Jane sat on the floor with Happy, her Australian shepherd, curled up next to her at our feet. We were so stunned—so frantic for answers—we did it all wrong. We didn't bother to toss pennies. And instead of asking a question, Jane just plopped the I Ching down and let the pages fall open. Hexagram 13: "Heaven over Fire. Fellowship. Harmony. Socializing. No person is an island."

Jane looked up at us bewildered, disappointment flooding her face. Her dark brown eyes flashed with anger. For years the I Ching has brought her grounding, order in an arbitrary world. Each random penny toss has consistently revealed some cosmic truth that reset the universe for the next rotations in her life.

Not on this catastrophic night. Camaraderie and harmony made no sense in our flame-filled world. We had lost an entire town—businesses and hundreds of houses. As far as we knew, Jon and I had lost our home as well. Jane, our evacuation host, was gobsmacked. "We want deliverance," she ranted into the void. "We want to be saved from fire." The I Ching had never failed her before.

I have always been skeptical about rituals promising panacea, so I dismissed the indescribable hexagram as an annoyance. What drew me were the open pages of the I Ching and the Chinese characters printed in heavy black ink: Heaven. Fire. Each was as recognizable to me as a favorite flower along a familiar path. Every Japanese schoolkid learns these *kanji* in the first grade. Their Japanese pronunciations floated through my head in murmurs soft as a lullaby. I pictured them combined with other *kanji* in words like *tenki* (weather) and *hibachi* (charcoal burner). I drifted into a dreamy escape, floating over rice fields terraced into steep hillsides. It lasted mere seconds before I jolted back to reality: Greenville was gone, reduced to ashes and rubble. My sweet little office at the top of the stairs—gone. Did our house burn? Our neighbors'? Did anyone die?

The universe has a way of bringing order to chaos obliquely and in its own time. We could not know then how patiently heaven would embrace us in steadfast encircling mountains while fire transformed our world. We could not know then how we would grow into the I Ching's elusive message.

Old Beginnings

I never really meant to live in this rugged logging and ranching outpost at the northern end of California's Sierra Nevada mountains. When I arrived in Chester in 1969, the ink was still drying on my Harvard Master of Arts diploma. Jon and I had come for a summer and had no plans to stay beyond that. This was to be another chapter in our adventure together, a journey that began in Japan, where I had been living for two years, and included hitchhiking through nineteen countries in Asia and Europe before flying back to New York. Two years at Harvard were enough stability for a while. So we drove across the country in a 1959 green Dodge pickup truck, which broke down in every state from Illinois to Idaho. When we arrived in Chester, we had no money and no place to stay. Jon had a summer job on a Forest Service helicopter fire-fighting crew, and I was blissfully without plans for the first time in my life.

I felt no more at home in Chester than I had in Kabul. My hair was too long and my skirts too short. All spring I had marched in protest against the Vietnam War, and for most of one long night I occupied an administration building in Harvard Yard. I was rebellious and slightly clueless, a young academic fluent in Japanese and well-schooled in Japan's cultural history. I had landed in a place where local sons were dying in Vietnam, and most people didn't know the difference between Japan and China.

I was as ignorant of them as they were of me. I didn't know what anyone meant when they referred to "the mill," the economic mainstay of this timber-dependent town. I thought it was stupid to call the mountain southwest of Lassen Peak "Brokeoff," tone-deaf to the more poetic cadence of the grammatical error. I scoffed at signs warning of deer crossings, knowing nothing about the habitual patterns of wildlife. I was a complete social misfit.

But the mountains were welcoming and the creeks inviting. I spent the summer swimming and basking on midstream rocks, hiking back-country trails, and exploring remote canyons flush with berries and wildflowers whose names I did not know. I was captivated by the towering trees and volcanic rocks so light I thought they might float. It didn't take long to learn that I had lucked into a unique geographic area, where the Sierra Nevada, Cascade, Great Basin, and California Central Valley bioregions jostle against one another. I fell hard for the Feather River watershed and its alpine lakes, geothermal fumaroles, and biological diversity. The hallowed halls and ivied walls of academia paled in comparison to the excitement of Plumas County.

Fire was part of the fascination. Even then I understood that it is essential to this landscape. Western scientists were just beginning to document the importance of the blazes that consume brush and small trees, cleansing the woods and leaving the large old-growth trees unharmed. Mountain Maidu, the Native Americans who have lived in Plumas County since their beginning of time, were way ahead. They had been using fire as a tool for millennia. They burned around their villages for protection from uncontrolled flames. Hunters tossed lit sticks along their paths to keep them open. Maidu also understood that fire renews the land, removing decayed plants and making room for new ones that attract wildlife. They used it to encourage the species essential to them as food and medicine. Hints about the complex role of fire in this landscape were part of the lure of Plumas County. And fire made Jon a good living.

So we stayed: for a winter, then another summer, then years. Indian Valley gave us a place to raise our two young sons among the wooded hills at the end of a paved road two miles from Greenville. We raced a team of Alaskan husky sled dogs in the winter. When the boys were old enough to carry backpacks, we three took to the trails of Lassen Volcanic National Park while Jon worked throughout the West as aerial supervisor of firefighting aircraft.

After stumbling into a job with the local weekly newspaper, I found myself hooked on journalism as another way to explore this landscape. I soon began freelancing, first for regional newspapers, then magazines, gradually building a career out of trying to understand the natural resources of this place and beyond, and the people who work to conserve where they live. My Japanese dictionaries gradually moved to back shelves, the *Manyoshu* and *Tale of Genji* replaced by books like *Cadillac Desert*, *The Forest Primeval*, and *The Song of the Dodo*. But I never put Japan away in boxes.

Journalism gave me a role in my adopted community, but it also kept me from belonging. I hid behind the objectivity of "he said" and "she said," clinging to the safety of separateness. I jumped into activities that involved my kids but stayed largely aloof from significant community involvement. I seldom named Greenville as my hometown, opting instead for the relative anonymity of Plumas County. Happy to hang out with family and a handful of close friends, Jon and I built our lives together around this place we both found so compelling.

The Dixie fire upended that life, pitching me out of the safe refuge I had created and into the unknown.

Evacuees

On August 5, 2021, the morning after Greenville burned, Jon and I left the haven of Jane's house in Quincy. We did not know if the flames that obliterated our gritty little working-class town had also taken our home of forty-six years. Everything pointed to utter disaster, so we joined our neighbors at a resource center the Red Cross had hastily set up. It was already crowded with people we had known for years. We followed them from one emergency relief table to another. United Way of California was offering a $500 Visa gift card requiring fourteen days to process. So much for immediate assistance. We shuffled past the California Franchise Tax Board table and the California FAIR Plan and Aegis insurance tables, part of a bedraggled line of evacuees facing uncertain futures.

We were at the county assessor's table when a red-code alert blared, first on one cell phone, then hundreds in ragged, heart-pounding succession. When we realized the emergency was in another county, our momentary dread lapsed into nervous giggles. No one panicked. We had been living with this fire for twenty-four days and were by now red-code veterans. As I watched my neighbors take in this latest alert with stoic patience, I felt an unfamiliar surge of affection. These are strong, self-reliant people: the handsome, big-boned woman who worked as a hairdresser in the brick-walled beauty salon on Main Street, likely the town's oldest building; the sweet man who shepherded a friend through hip replacement with doctors a hundred miles away; even the skinny elderly man who has been yelling at me since 2005 in an incoherent alcoholic rage.

Greenville is a hodge-podge of loggers and ranchers, rednecks and hippies, Mountain Maidu and retirees and deadbeats. The town has been in physical decline since I moved there, slowly sinking into its foundations. Those of us who have hung onto it are stubbornly strong-willed. We seldom agree about politics, religion, or the weather. We squabble over water ditches and which civic organization's insurance should cover which parade. We picked to death a solar-power project and took five years to decide on a design for a community building. For most people, high school football is holy and climate change a hoax. We are obstinate; feuds over kindergarten Valentine cards can last a lifetime. Most days we shared this place in uneasy coexistence, a community of the doggedly independent. But bring in an outside adversary with the odds stacked against us and all that defiance becomes a united force to reckon with. Now we faced a threat like none ever.

Climate change is the disaster lurking for all of us. Fire delivers it to some; floods, drought, and famine to others. The Dixie Fire brought climate change to Greenville with a vengeance. Sparked by faulty Pacific Gas and Electric equipment on a power network forty miles away, flames had been moving up the Feather River Canyon since July 13. They met with overstocked forests mismanaged for a century and dried crisp on a warming planet. Eleven days later Dixie stormed into neighboring Indian Falls, burning nine out of thirty-two houses in a pyre that sent so much black smoke into the surrounding valley that the crickets began humming at four o'clock in the afternoon. Ten days after that, a forty-thousand-foot pyro-cumulous cloud the color of bruised flesh collapsed over the ridge south of Greenville. Soon flaming branches and red-hot embers were hurtling down the mountainside, torching trees as fire roared into town.

We watched in horror from Jane Chang's house, snatching grotesque images from Facebook, chasing down Twitter links, and trying to make sense of the devastation evolving before our eyes on infrared maps. It took less than thirty minutes to reduce the town's tarnished Gold Rush charm to rubble. The Dixie Fire raged for more than three months through nearly one million mostly forested acres—from the Feather River Canyon to Lassen Park, the distance from San Francisco to Sacramento. It devastated the community of Canyon Dam and destroyed nearly one hundred homes in Warner Valley along with Greenville and Indian Falls.

How do you live in a town with no post office, no hardware or drug store, no library or gas station? How do you recover from a fire that decimates more than one thousand homes? Why would anyone want to live here again? None of us at the Red Cross center had answers, and neither did the Red Cross. Yet even on that day of reckoning, a day after utter destruction, I knew that Greenville would mount the fight of its communal life. I watched a woman gently guide her elderly mother from one resource table to another.

A young father held tightly to his daughter's hand as he gathered up application forms. I was overwhelmed by an unexpected, urgent commitment to these people, this town. We were equally bereft, facing a shared future equally bewildered. Despite my ambivalence over belonging, I knew that these defiant loners would overcome these odds as a community. I had a surprising urge to join them.

Asking Jane Chang to let us evacuate to her house was an impromptu decision. We knew one another in the vague way that like-minded people in neighboring towns sense compatibility. And we'd recently taken a road trip together. Jane was raised by a mother who survived the fire-bombings of Tokyo and a father who survived the atom bomb that annihilated Hiroshima. Our trip was to interview him. Now ninety-four, he shared clear memories in vivid detail, some new to Jane. She and I brought different perspectives to these conversations, but we were equally engaged. It was a promising friendship. Still, as I spoke to her on my cell phone in smoke-choked Quincy after our mandatory evacuation from Greenville, I could not have guessed how serendipitous a choice it would be.

Jane's house is down a few steps from the street in a quiet section of Quincy, the Plumas County seat. When we arrived, a stand-up bass dominated a wall in the living room, but my eyes went straight to the Butsudan hanging above it. Every Japanese Buddhist household has one to pay respects to the Buddha and honor family members who have departed. Jane's Butsudan is the size of a bathroom cabinet, with delicate gilded doors opening to an ornate platform. A Kannon Buddha stands before a porcelain bowl for food offerings. Jane has added several small statues of Buddhas with particular meaning for her. With this Butsudan, and with Japanese bowls in a kitchen stocked with foods familiar to me from the Japanese homes I had lived in, I immediately felt welcomed.

On the fateful night that destroyed Greenville, we were still processing our puzzling encounter with the I Ching when Jeff LaMattina, Jane's partner, burst in with Ruby, his protective dog. He had stayed in the heat of the fire as long as he dared, trying to defend his house on a hill above town. He knew he had likely lost it by the time he dashed through the flames to survival. Hope is an awkward ally, but we clung to it that night. Dixie had brought us together, two couples and two dogs seeking order in a capricious universe. For Jeff, confirmation of disaster arrived with abrupt cruelty the next morning. The house he had built was gone. He poured his grief into a frenzy of activity, bent on rebuilding as swiftly as possible.

Jon and I had a longer wait. Using sketchy internet maps, we anxiously traced the line of flames advancing east, west, and north from burned-out Greenville. Our property on Pecks Valley Road was in an ambiguous zone. The town of Greenville was obviously inside the black fire line, clearly scorched, but just what had burned on our road and in the surrounding forest was unclear for thirty-eight uneasy hours. Dixie forged an erratic path, circling back through forests overcrowded by a century without the benefit of natural fire and overheated by a planetwide 1.9-degree Fahrenheit temperature increase.

Mid-morning on August 6, Jon and I were alone with Happy in Jane's house when we got the phone call. Our place had been spared. Our house, the timber-frame barn Jon built from our own trees, his shop, our son's cabin—all had survived. Jon and I hugged each other and danced for joy, with Happy spinning around us as we rejoiced under the protective gaze of Jane's ancestors and the Buddhist saints enounced in the Butsudan. We called our sons, laughing and sobbing. It was a bittersweet moment. Dixie's chaotic romp through our valley grabbed some homes for annihilation and randomly bypassed others. It claimed Jeff's, sidestepped ours. There is no logic in a climate disaster, no order to its destruction. Jane Chang's evacuation haven was now a household of arbitrarily selected victims and survivors, a microcosm of disasters everywhere.

Over the next three weeks, Jon and I adopted a routine designed to give Jane solo use of as much of her own space as possible. We sought nearby out-of-the-way places for solitude and solace. One of them, Dellinger Pond, was close to Jane's house. The trailhead sign is all but hidden and the trail an overgrown tangle of blackberry canes, star thistle, and other invasive plants. I sometimes left Jon in the company of Happy and walked there alone. It became my refuge, a place to complain to the universe about its callousness. My rants released the most guttural utterances I have ever heard coming from my own throat—raw moans I did not recognize. I was angry, yes, but also overwhelmed by unspeakable despair, with no one to blame save PG&E. Despite the crimes it has committed against community after community, I knew PG&E was not the true object of my anger. That was climate change: the wanton excesses, abuses, and overuses of natural resources our society has indulged in for generations. The hotter, thirstier planet is drying out the forests that protect us, and the precipitous rise in temperature is something from which we cannot recover. And I knew too that it was simply change I protested: the uncertainty, the instability, the great unravelling that has torn apart a human community as well, flinging its fragments to the four winds to find their way alone. I was scared.

One afternoon, after a particularly thunderous rant at Dellinger Pond, I sat quietly on the ground, savoring the calm that always follows the release. I was gathering myself to reenter society when a movement caught my eye. A pair of sandhill cranes was grazing in what by August was more marsh than pond. They must have been there all along. Transfixed, I watched their red crowns bow and dip as they moved through the reeds and sedges in an ancient, delicate dance of survival. Soon a coyote emerged from the far side of the pond, ambling along the shoreline away from the sandhills before disappearing into the trees. And Dellinger had one more gift: a fox. She trotted within five feet of where I stood, looked me in the eye, and moved on. I breathed more deeply than I had in weeks, gratified by this reminder of the scale of the universe and my small place in it.

That night Jane made *zaru-soba*, a cold buckwheat noodle dish that is one of my favorites. While cooking she tossed out Japanese phrases from her childhood, sometimes spontaneously bursting into song from the days when Japanese was her only language. I laughed at her baby words, but she used phrases I did not know—sassier than what I had been taught, more endearing. At the oddest of times, in the most unanticipated way, I was reclaiming my Japanese background. This evacuation host so generous with her time and space had even more to give me.

We endured a full five weeks of evacuation before we were allowed to return to Pecks Valley Road. On our last day at home before we left—that lifetime ago—we had seen a pair of white-headed woodpeckers on the ponderosa pine outside our living room window. A three-point buck lounged just outside the garden fence. The rabbit we saw occasionally also made an appearance that day, hopping through the smoke as we packed up to leave. Despite our nervous haste, I kept thinking about *The Once and Future King*—the way the animals of the Wart's childhood showed up to give him strength as he labored to pull the sword from the stone and claim the kingdom as King Arthur. On my last look behind the house I glimpsed an enormous bird with a distinctive red crest in an oak tree. Could a pileated woodpecker have graced our evacuation departure?

Now we were limping home, anxious about whether any of the wildlife had survived and how we would find our house. It loomed through the heavy smoke, smaller than I remembered, grimy but intact. My heart raced. Home! Still standing! A layer of greasy ash coated every surface. We had water from the well but it fluctuated between fire-hose surges and drizzles. Electricity came from PG&E generators belching diesel at the end of the road. We had no landlines, no internet, and patchy cell phone service. No gas station, grocery, or hardware store. No town. Our garden was dead. As fortunate as we felt to have a home, this was no celebratory return. Yet somehow the wildlife survived. The buck was even bulkier, carrying his rack proudly as he paraded past the house. The rabbit returned—likely never having left. We did not see the pileated woodpecker, but a resonant hammering from deep in the woods told us he was about.

In the weeks that followed, all of us lucky enough to have homes at all toiled in the ash that worked its way into our lives. The Dixie Fire unraveled us. It left enough strands in place to pay the bills, put food on the table, and respond to well-meant inquiries from family and friends. *Yes, thanks, we are doing fine. No, our house didn't burn. Yes, we are so very fortunate.* But at night, every night, we were left with the holes between the few remaining threads somehow holding us together. Where would we find replacement strands? How would we summon the skills to darn them into the tatters? Did we even want to? It was a shared quandary. Each of us claiming Greenville as home—even the ambivalent—has contributed to the warp and weft of the community. Now it was badly frayed.

My dreams careened: In one sweet scene a local couple is dancing cheek to cheek celebrating their forty-second wedding anniversary on the floor of the high school gym. In another I leave a dishrag on the coils of an electric burner, and the smell of singed cloth wakes me up in a panic. Other nights pages of burned books drift through my dreams, imprinted with faces: a blue-eyed woman with a voice like a red-code alert, a grocery store clerk with straight black hair cascading down his back. We lock eyes before they sink into the dark.

Sleep became a communal issue. Conversations before the fire focused on the latest lapse of the Plumas County Board of Supervisor and annoying garbage bears. Now they were about the relative benefits of melatonin, chamomile extract, CBD gummies, or just plain pot.

The Bowl

One day in early September, my officemate of eleven years came to the house. She and her husband had been sifting, a sordid act of hope that by scratching through the ashes—the rubble, the melted plastics and other hazardous materials remaining from our office—something of value would emerge. I didn't have the heart to do it. Instead, I haunted the ruins, standing at the edge of the pavement below what had been the double windows of my office at the top of the stairs. I stared into the grimy detritus of my forty-year journalism career. I hoped the sun would produce a gleam, a refraction off the inch-thick engraved glass award presented by the Society of Environmental Journalists in 2014 for the best feature story. It was about Fukushima and Japan, nuclear disasters far worse than the Dixie Fire. Glass melts, I reasoned, so perhaps some reshaped glob might have survived. Other times I focused on the rusting file cabinets leaning one against the other like fallen buddies. I didn't know which were mine and which my officemate's. I envisioned tugging a drawer open and finding intact the Japan file with all the photos of my Japanese family in Wakayama, where I had lived for two years. Or the Galapagos file and my notes on using selective breeding to recover the Floreana Island animal species, extinct since Darwin's visit in 1835.

But I did not sift. My officemate and her husband did. They found a precious Wedgewood vase that had belonged to her grandmother. My other officemate also sifted. She found a tiny ceramic vase she kept on her desk for paperclips. It was still full. I did not sift. Something about preferring my unembellished memories to twisted shards of objects I loved.

Others did what I could not. On that September day my officemate presented me with a small ceramic bowl she had uncovered. I held it in the palm of my left hand, turning it clockwise from the rim in an instinctive ritual learned decades earlier while studying Japanese tea ceremony. Pea-sized chunks of ash were fused to the shoulders; a tiny collection in one place looked like a miniature celestial constellation. Near the rim a bit of blue winked out from a splotch of grainy white. On one flank the dark brown glaze was shaded pale gray, lightening to a soft gold like the sky opening after a rainstorm. I turned the bowl to look at the base, honoring the late potter Paul Herman, whose name was still clear in the etched clay. I peered inside. White ash flowed down from the rim in soft clouds rounded toward the bottom.

The bowl, which had held my morning miso soup and lunchtime yogurt, was transformed. Its smooth sheen was gone, replaced by a rough texture that was oddly reassuring. The fire had taken a piece of functional art, drawn from its caldron, and created a darker, more complex beauty.

I was flooded with a surge of emotions so powerful I felt unsteady on my feet. The touch of each fire-fused mar transported me back fifty-two years to a treasure trove on the seventh floor of a warehouse on the Lower East Side of Manhattan. I had flown from

Boston to meet my Harvard professor, who was inventorying a private collection of ancient Japanese pottery. He asked me to help identify the pots. My major paper for my master's degree was a study of two sixteenth-century kilns near Nara, the ancient capital of Japan. Of all the world's cities I have visited, Nara and nearby Kyoto are my favorites. They were both an easy day-trip by train from Wakayama. On days when I was not teaching English at the university, I would hop the milk train for these historic cities, haunting the temples, museums, and ceremonial teahouses, absorbing the *shibui* elegance and mystery of a culture a millennium older than mine. The Zen arts were the most compelling for their simplicity, the way they incorporated natural materials and energy, and their emphasis on the process of creating, not the product itself.

At Harvard I had intended to study literature but was so drawn to these ceramic artifacts that I deflected from a field marginally practical as an occupation to one clearly useless. I happily toiled away in the basement of the Yen Ching Library, poring over dusty Japanese texts to coax out the history of the kilns that produced the pots most intriguing to me. Influenced by the Zen spirit of spontaneity and serendipity, the Iga and Shigaraki potters formed their bowls with local clays whose natural impurities yielded the unexpected. Once fired in wood-fired kilns, the pots took on ash and the colors of flames: a flash of iridescence here, a fleck of deep mauve there. I labored to translate the *kanji* that held their secrets. And in that New York warehouse I finally got to touch the rough-fired surfaces of these simple pots.

I didn't expect anyone in Plumas County to share my exuberance for this arcane aesthetic. And although I taught Japanese cultural history and the Zen arts to students at Feather River College and Lassen College, few did. So I held my pottery passion within and held myself apart, ever the misfit, if only in my own eyes. Now I was holding a bowl fired by wood in a local kiln and refired by Dixie in my very own office. Dixie took everything else: my Harvard diploma; hundreds of books in two languages; small clay figures salvaged from tiny shops in remote mountain towns; my original Kenkyusha Japanese-English dictionary and all the others so critical to unlocking the processes of those ancient potters. But the fire, so cruelly destructive to so many, had bequeathed this small treasure, a relic of my academic past forged in my traumatic present. It connected my long-siloed selves.

This was too much to explain even to the officemate who understood me well. So I simply thanked her and her husband.

In the months that followed the Dixie Fire, I slowly reassembled my life as a journalist. One of my sons had built a cabin when he was fourteen. I converted it to an office, replacing the bed with a desk. My first decoration was a four-foot Japanese hanging scroll with luminous charcoal-ink clouds hovering in an undefined sky above jagged mountains. At the bottom, all but hidden among rocks and trees, are three tiny huts. This has always represented my view of how humans fit into the world. Now this landscape painting joined my twice-fired pot.

The first post-Dixie book I bought was a Japanese dictionary. I thought it was an odd choice—less momentous but just as serendipitous as choosing Jane as the first person to ask if we could evacuate to her house. Along with the Philip Hyde photographs and Japanese woodblock prints, I most mourned the loss of my books. In conversation after conversation, I would instinctively reach for a specific favorite, knowing exactly where it was on one of my six sets of bookshelves. I could feel it there, the way an amputee feels an itchy foot. Each rush of sadness that came with remembering reality was a small stab in the heart, over and over for each book I reached for. The new dictionary was an ordinary Random House paperback. But having it where I could grab it, hold it, and open it was satisfying beyond anything I can explain. It was a new beginning.

Opportunity

Life in the Dixie burn scar was a combination of deep isolation punctuated by moments of intense connection. All of us able to live in homes close to Greenville spent hours on the road driving for the groceries, gas, hardware, and mail no longer available in our town. When the always shaky internet went out altogether, we drove to find it. We chased cell phone signals from hilltop to valley cove. Each trip involved long traffic stops for tree removal along state highways 89 and 70. Each stop added the insult of thirty-minute waits to the heartbreak of seeing century-old trees—gorgeous pines, stately Douglas firs, and cedars—dropped to the dirt and ground into chips before our eyes. We were witnessing willful devastation in the name of safety for the motoring public. Along with so much more, we were losing the trees that softened the roadsides and sheltered us from the steep slopes dropping into canyons.

Returning from Quincy late one afternoon, I was mourning these losses as I crested the last hill before the highway dips into Greenville. Suddenly Indian Valley emerged before me, a verdant, sunlit vista of pastures dotted with cattle and irrigated fields stretching to the slopes of the mountains. The peaks of Keddie Ridge rose above, stoic, enduring, and comforting. A rush of gratitude raced through me. I know this valley; I have climbed most of these peaks. I spent a summer as a lookout on this mountain. But never in my half-century-plus in Plumas County have I seen this view from this place. Dixie and the ravages of roadside clearcutting had bestowed us this gem, this silver lining of disaster.

In my elation I had flashes of my childhood: a small blond girl in a Dutch-boy haircut standing before an audience beside my older brother and sister, reciting the 121st Psalm: "I will lift up mine eyes unto the hills, from whence cometh my help." I live at the base of Keddie Ridge and have often sought its help. It has been a touchstone, grounding and reassuring me with its looming, reliable presence. The tips of the peaks form the profile of a prone man: an eyebrow above a large jutting nose, the hint of a jaw, and beyond, a bulging belly and toes. The Mountain Maidu call Keddie Ridge Chiwitbem Yamaninom. After Worldmaker completed his global tasks, he returned to Indian Valley to rest. According to one Maidu legend, he will sleep until the last Anglos leave the valley. According to another, when Worldmaker arises it will be the end of our time on Earth.

After the numbing isolation of trying to restore our lives in a burned-out town, the announcement of a mid-November townhall meeting was as welcome as a party invitation. Nearly one hundred people gathered in the elementary school gym, where most of us had watched our kids perform plays like *The Velveteen Rabbit* and *Who's on First*, where we heard the first rumblings of music from trumpets and trombones, where we had eaten far too many tasteless lunches out of unconditional love for our children. Now we heard from community leaders about the official recovery work underway by state, federal, and nongovernmental organizations. It's the beginning of a long road, they told us—a really long road that promises gas stations, food trucks, fiber-optic cables, and houses . . . eventually . . . if we all work together.

I looked around the room filled with ranchers who had defended their homes against the Dixie Fire, schoolteachers who had lost theirs, store owners whose businesses were rubble. I expected an onslaught of accusations over who didn't do what when and why not. I expected carping familiar from decades of attending similar townhall gatherings. The usual suspects were there: the aging logger who had never gotten over the limits placed on clearcutting, the business owner who made fun of everyone who didn't look like him. Harmony in this crowd? When pigs fly, I thought. I steeled myself against the finger-pointing.

What I heard instead was hope born of gut-wrenching grief. I felt tenderness amid trauma and an unspoken pledge to cooperate that left me light-headed. *These are people committed to rebuilding*, I thought. Maybe we can do this together. It was a Mountain Maidu tribal leader who provided the perspective most of us lacked. She welcomed us to Kotasi, our little town "under the snow line." She reminded us of past disasters visited on Native Americans, "the destruction of our community time and time again." She spared us the shame of hearing her name the destroyers, instead gathering us together in the possibilities of this moment of renewal. We have been offered an enormous opportunity, she said. "We are all rediscovering our relationship with place."

Climate disasters choose their victims indiscriminately. How we survive depends on our willingness to accept change. Dixie was forcing me to surrender my self-imposed isolation. It was forcing Greenville to embrace an uncertain future. When a woods-hardened man stood before us at that first community meeting, voice cracking with anguish, the room hushed. How could I resist this summons to join what was beginning to feel like a community I could commit to?

Throughout the fall I felt the wrench of my ruined office. It had always been more than a place to work—it was a room of my own, and I adorned it with personal tchotchkes gathered from journalism forays to faraway places. The universe had spared me the horror of losing my home, but this loss was painful. I haunted the wreckage like a widow at a gravesite. I sang strains of Gustav Mahler's Symphony No. 5 to the ashes, laughing at myself when I wasn't crying. I photographed crows in the black locust tree, charred but regal, still standing outside what had been my bathroom window. I cursed the crew that finally cut and hauled it away. Dixie delivered a gut punch on every drive through Greenville. It hurt to see the crooked angles on metal roofs now stoved-over kitchens where papas cooked breakfasts for their kids. I recited their names as I passed homes reduced to hovels.

Christmas Eve snows softened the raw hulks of the houses, highlighting the tiny colored lights people had strung around stumps and along walkways that now led to nowhere. As more snow accumulated, the town took on a festive look. No one was rejoicing, but we welcomed this fleeting beauty. Spring brought daffodils planted lovingly in yards no longer there. Hollyhocks bloomed in outrageous shades of optimistic pink. Spring also brought demolition crews, scrappers, and haulers in an endless parade of gigantic backhoes, cranes, and dump trucks. They leveled the angular mounds of Greenville's eight hundred lost homes, leaving anonymous lots with dirt as fresh as new burial plots. I anguished over the realization that it was becoming harder and harder to remember who had lived where.

Townhall meetings were now monthly, bringing the community together in increasing numbers. Leaders of all ages were emerging to help meet the needs of people who still lacked permanent housing and to generally insist to FEMA and other bureaucracies that we would not accept cookie-cutter solutions. This is Greenville, still gritty but newly united in an odd but determined alliance. Some people reported on house plans preapproved by county officials; others shared efforts to restore internet access and the promise of a pop-up business district. What had been vague, months-long hope became reality as the first 2x6 uprights jutted into the June air, the first new homes for families.

Spring also brought renewal to roadside clearcuts. Oak stumps sprouted, decorating the slopes with bouquets of vibrant, defiant green. The flames that left Greenville in ashes were rejuvenating the highway corridors in slopes painted orange with California poppies. Deeper in the forest, Dixie was transforming overstocked stands of white fir, cedar, and pine, disrupting the stability that suppressed new plants and wildlife. Fire takes away old-growth habitat for spotted owls and Pacific fishers, but it stimulates habitat for mushrooms, insects, and birds—an entire suite of species that has evolved with fire. I have stomped around stumps still burning from the previous year's fire to witness a brilliant blue invasion of lazuli buntings and hillsides knee-deep in silver lupines. But Dixie was the largest single fire in the state's recorded history, driven by a climate that is changing before our eyes. It had new and troubling lessons to teach us.

Echo Lake

A year to the day from our second evacuation, I headed to Lassen Volcanic National Park. I was desperate for the solace of time alone in the outdoors, the only place I have ever found truly healing.

The trail to Echo Lake starts with a wooden boardwalk crossing a marsh before leading directly into a stand of burned trees. The Dixie Fire blazed through 68 percent of this park. Here the burn was moderate, sending flames ten-feet long up the trunks of red firs and mountain hemlock, claiming some but not all of them. It left stump holes as big as VW bugs, with hollowed tunnels where roots had kept their trees alive. Sagebrush lizards scrambled away from the gigantic shadow I cast on burned-out stumps, hiding among the splinters that must have seemed as big as cliffs to them. After a winter of needle fall and snow, the soil beneath the trees was colored like a crown of dark hair streaked with henna.

Elsewhere, however, the Dixie Fire was so intense that it killed acre after acre of old-growth trees. At just under a million acres, this was no proverbial cleansing burn simply restoring essential flushes of nutrients and bursts of sunshine. Dixie took the fundamental processes of ecosystem disruption and rebirth to an extreme neither Mountain Maidu nor Western fire scientists had witnessed. After a century of fire suppression and management emphasizing old-growth logging, the forests left in its path were overcrowded with small trees and brush primed for a runaway inferno. With climate change warming and drying these forests, Dixie upended what scientists thought they knew about fire behavior. It left the forests I know and love the best in ecosystem chaos.

I followed the trail to Echo Lake up a flow of lava from the eruption of Hat Mountain, one of the park's cinder cones. Sparse stands of conifers baked in the hot July air. Clark's nutcrackers were the liveliest sign of life, sending their metallic squawks across the mountains. I was more than ready when I spied a sparkle of blue flashing beyond the trees. Flames had licked right down to the shoreline, scorching logs that jut ten feet into the water.

I settled in the shade near a spit of sand next to a patch of Brewer's mountain heather covered with fuzzy pink blossoms. The lake was azure, the surface glassy. A bleached buckskin log protruding into the lake was just too tempting. I stripped, slid along the smooth surface gingerly, and then took the plunge. The water was deliciously cool. I rolled onto my back into a float that transported me far from the Dixie Fire. Clouds drifted across the sky like beings with intention but without hurry. I envied their tranquility. Some had the shape of the four winds in my childhood picture books: wild, streaming, Einstein hair and friendly faces. I closed my eyes, resting in a space so serene I forgot time and place, rage and alienation. Long minutes passed. The heavens held me in an embrace anchored by the surrounding mountains. I felt safe. When I opened my eyes, I burst into giggles. No doubt on this July day I was the oldest skinny-dipper in Lassen Park.

I dried on the log in the sun, sedges tickling my toes. Some had inch-long pods in bright green and brown herringbone patterns. I was still sitting on the log when the wind began to stir the water, lapping it against the log like smacky kisses. Soon a few raindrops fell: blessed summer rain. The drops hit the placid lake surface in tinkles that echoed across the water, the gentlest of percussion choruses. Then they increased with a crescendo that became a full-fledged storm, raindrops as big as hailstones. I moved to the shelter of trees, leaning against a rock that held the warmth of the sun now eclipsed by dark clouds. By the time the rain stopped I was fully soaked and as deeply calm as I had been in months.

I left Echo Lake reluctantly, hiking back through puddles and grasses refreshed by the rain. As I came through the burn and rounded a final corner I was confounded by the smell of diesel and the hum of a generator. The campground near the trailhead was full of people enjoying the park, blissfully unaware of the Dixie Fire, of climate disaster or communal loss. I burst into tears. Change in the midst of catastrophe requires more strength than I often had. I was as vulnerable as the blossoms of silver lupine shimmering along the trail.

Uneasy Harmony

Greenville has never been a place to take hardship lying down. On the first anniversary of its demise, Pine Street hopped with food trucks lined up across from where the historic Methodist Church had stood. Burgers, bratwurst, and bowls of sautéed meats and vegetables were served. Bartenders poured beer, wine, and mixed drinks from a truck that will have to make do until the new owners can rebuild The Way Station, Greenville's only bar. We called the temporary digs The Way Baby. Where charred ruins had lain, Greenville now had The Spot, a temporary business district and proof of the community's resolve to rebuild.

Beyond The Spot, the townsite stretched out to meadow and denuded forests, their trees still upright like haunted ebony specters. Few of us will live to see stands vibrant with Western tanagers flitting among the eighty-foot crowns of Douglas firs. We will not lie on spice-scented cushions of matted needles to stare at the sky through pine branches filtering sunlight. But crisp September nights will soon tint the understory with crimson shrub oaks, and autumn rains promise morels. Forests that have evolved with fire will return, perhaps with more resilience than those we lost.

On this August anniversary day I clinked plastic beer cups with a woman who is planning a community garden next to where the three-story hotel once stood, with a soft-spoken Maidu man who vows to rebuild the center where children gather to learn the traditional language of this land. The Dixie disaster pitched us into the unknown, forcing us to find our own way out of the chaos. Nothing is yet safe, no one unchanged. But the promise of the I Ching hovers over recovery, Greenville's and mine. Hints of the fellowship once so elusive are burgeoning. We know salvation is not heaven-sent. Greenville is evolving house by house, soul by soul, in a profoundly human effort driven by guts and communal will. It is a disorderly process of false starts, failures, and small successes. If the harmony that has arisen serendipitously from the ashes endures, we may yet build a community that honors its past and embraces its future. I may yet be a part of it. The I Ching didn't say when.

Formerly Incarcerated Firefighters as Community Servants
An Interview with Brandon Smith of the Forestry and Fire Recruitment Program
Dani Burlison

Brandon Smith never knew he could be a wildland firefighter, and it took him being impacted by the justice system to find out that he could. Today he is the cofounder (with Royal Ramey) of the Forestry and Fire Recruitment Program (FFRP), which has locations in Oakland and Southern California. A formerly incarcerated firefighter who took part in the inmate firefighter program, Smith faced several challenges securing employment once released from prison. He founded the organization to train others like him—formerly incarcerated, mostly Black and Latinx people—who had training in prison that wasn't necessarily recognized by firefighting agencies.

Incarcerated firefighters make up about one-third of California's firefighting force, and earn roughly $2–5 per day, with an additional $1–2 per hour while serving on an active fire. The work is grueling and includes long days on the frontlines of wildfires, primarily clearing heavy brush and cutting firebreaks. Despite the extensive training and experience they gain from Cal Fire while in the state correctional conservation camps or "fire camps," many face barriers to employment in the field once released. One issue is that many agencies require firefighters to become certified as EMTs (emergency medical technicians), which is currently not possible for those with two felony convictions on their record in California. For those with one felony, many do not become eligible for EMT training and certification until five to ten years after their sentence ends, depending on what type of conviction they have.

In 2020, California governor Gavin Newsom signed bill AB 2147 into law, which would ensure that formerly incarcerated firefighters who worked in California Department of Corrections fire camps without any issues could apply to have their felony convictions expunged. However, it can take up to one year or more for records to be expunged, and not everyone is successful with their petitions. Advocates say AB 2147 isn't the miracle bill many had hoped for. With a firefighter shortage—the force fluctuates between 650 and 2,000 workers in California—hiring from the large pool of available and trained formerly incarcerated firefighters seems like an easy solution.

According to the California Department of Corrections and Rehabilitation, inmate firefighters receive just four days of classroom training and four days of field training before they join crews battling California's deadly wildfires. Despite the extensive experience during active fires, there remain gaps between hands-on experience while in California prison system's fire camps and employment opportunities upon release. The Forestry and Fire Recruitment Program helps participants bridge those gaps through paid training, mentoring, advocacy, job coaching, and connections to employment. Brandon currently lives in Altadena, California, near the Angeles National Forest. We spoke via Zoom for this interview about his work.

DB: I'm curious about how you first got involved with the inmate firefighter program. What was that like?

Brandon Smith: It was scary. At first, I said no. I remember I was in a cell, I was at Wasco or Delano State Prison, and the woman came to my cell, and she asked me if I would be interested in joining the firefighter program that CDCR [California Department of Corrections and Rehabilitation] had. I said, no, I didn't want to be a firefighter. Fire was not something that I wanted to be a part of.

And I remember, when I said no, there were so many people in my facility who were like, "Bro, you said no? You said no? You're one of the only ones who can go out there." And I learned it was going to be a great opportunity, and I chose to say yes. It was a great moment in my life, and it changed my trajectory and the trajectory of my family moving forward.

DB: What was it like to go out on your first fire?

Brandon Smith: When anybody hops into a dangerous situation, the first thing that hops up is adrenaline. So that's what came into my body. I remember hopping out of the fire engine with adrenaline running through my body, because I did not want to be there. I was praying that the Lord could save me. And when I hopped out, it literally was like an *Avengers* movie. You got things flying, you got machines, you got strong people, you got everybody moving in multiple different aspects, making things happen. We went out there, and we came together for the common good.

Fire has been one of the most physical experiences that I've ever had in my life, and one of the most rewarding experiences I've had in my life. We went out there to go help people, and we did. And coming into realization of that the morning after was extremely beneficial. Because I came from a place where I was considered a public nuisance. And then to do all this work that night and wake up in the morning and find out that we saved this town, I was just... *I can be a public safety officer; I can be a public servant. I can go help out for the greater good.*

DB: I have this assumption, and I think you kind of touched on it, that fighting fires really bonds people together. Has that been your experience?

Brandon Smith: It was great coming together and being able to learn and identify things with each other. It didn't matter where we were from, it didn't matter what race we were, it didn't matter what area of Los Angeles County with gang members we were from. It didn't matter, none of that. But we all came together for the very common good. Being able to learn together, to grow together, to work together was extremely, extremely helpful. And it was a blessing, because we knew that we could all come together. If a place is on fire, I don't care where you're from, we're about to go save it. We're about to go fix it right now. And that's what we started to learn.

DB: Clearly this had a big impact on you. What is it about this specific kind of work with fires and other firefighters that led you into this life path once you got out of prison?

Brandon Smith: There's something that's very cleansing around fire. And in my line of work there is an opportunity to bring the best out of folks and then also cleanse, and/or figure out folks' transgressions. A lot of justice-impacted people are actually community servants, and they just don't know it, and/or understand it, and/or the news or the community portrays them in different ways. And fighting fires is a way for us to clearly go show and say, "Hey, I'm here for my people, and I'm going to help people. I'm going to help folks out. I'm a community servant."

There are not that many people that look like me, come from where I come from, that have natural access to these careers. And we at FFRP, our whole goal is to make sure to go say, "Hey, you can go do this work. You are worthy." A lot of folks think of us as public nuisances, and we are public servants. So we just work to go support them.

DB: When you left prison, did you immediately want to work as a firefighter? What was that process like?

Brandon Smith: Yeah. While I was incarcerated, as a firefighter with the State of California, I grew to love the career. I grew to love the career so much so that even before I was released, I had asked folks "How do I do this job professionally?" Nobody knew. The fire chief didn't know, the correctional—the COs, they didn't know. So it was all of us trying to figure it out, like, "Oh, well, Brandon wants to do it, but none of us know how it's going to happen." I came home, I spent two years trying to navigate the pathway. I would apply to fire stations, I would apply to fire departments, and I would get continuously denied. "Oh, you don't have the qualifications, you don't have the certifications, you don't have the experience." And I'm, like, "Well, wait, I just been doing this."

I found a very unique one-off way to go hop into this work. The chief of the Big Bear Fire Department gave me my first job because she saw that I could do the work. How she saw that I could do the work is I ended up going to a firefighter training academy, and I graduated top of the class of the academy. And everybody's like, "Wow, you graduated top of the class." I'm, like, "It's not a big feat for me, bro, because I've been doing this for two or three years."

When I graduated, she came to me and said, "You know what, let me give you a shot. Here's the application, turn it in directly to me." And that was a pathway for me to get into this work, after two and a half years of really trying to do this. It was so great. I had felt myself as a—maybe the term is an outlier or special or something like that. I was, like, "I'm the only one that's been formally incarcerated, from fire camp, doing this work professionally."

And then, though, while I'm working on a fire, literally in Big Bear, California, I hear somebody calling my name. And I look, and I go see, and it's all the people I was formally incarcerated with three years ago.

"Wait, what? Wait, you a firefighter? How did this happen?" And with that is where we decided to go start FFRP, because it was, like, "I cannot be the only person out here with this experience. We need to go be able to go support other folks."

DB: What was that process like, starting the organization?

Brandon Smith: It was extremely hard. We started in 2018 as a group of volunteers, myself, and my partner, Royal Ramey. We started the organization. We were firefighters in the San Bernardino National

Forest, and with Cal Fire. And we were just putting our own paychecks into this to make it happen. We did a lot of research, like: how do you start a new nonprofit? What are the structures? How do you fund-raise? We Google-searched everything.

And then what was interesting is that two people came into our lives that really transitioned the direct work that we do. One was actually Google. Google was, like, "Wait, I see you researching something out here. What y'all talking about?" So we started talking to Google, and Google supported us. It's crazy. It's kind of backwards, right? It's kind of weird.

Then the second one was John Legend. He came out here and was a really big vocal supporter of the work that we do. And we started to grow and work and push forward. Ever since then we've been focused on program development and growing our groups and supporting more folks.

DB: How did things progress once you got some funding? What kind of community started developing?

Brandon Smith: Our first year we graduated eleven people; they all have careers. Our second year we graduated thirty-three people; thirty-one of them have careers. One went to EMT school. One of them decided to move out of state. Last year we graduated thirty-two people. This year [2023] we're going to graduate fifty people.

And we've expanded. Now we have an office in Pasadena, California, the Inland Empire—San Bernardino, and we also have an office in Oakland. But the whole thing is, there's a need for people in this space, and there are large communities who are not aware of this career opportunity, and we're just trying to share that with them.

Our goal is to get more women into this space, more people of color, more marginalized communities. And so FFRP comes at this very unique intersection of multiple needs and concerns.

We at FFRP are actively trying to recruit and support as many women as possible. One of the things that I've seen moving forward is that women are feeling more emboldened, so we at FFRP are constantly trying to support that. I will say, even though more women are trying to go hop into this space, there are still multiple challenges for women.... This sector is led by men, by white men, by white men from non-urban communities.

And the majority of women and folks that we deal with are women from marginalized communities, in urban communities. So it's the exact opposite, and there are challenges. One thing that we've been committed to at FFRP, and actually a couple of funders, is that we are trying to set up our own twenty-women fire group.

Our challenge right now is just trying to find the women to go hop into that space. For example, between our SoCal and Bay Area groups, we have five women that are moving forward, and we're trying to help them out. And hopefully we get lessons learned on how to best support them and go from there.

I recognize that I come from a male-dominated world, and I'm a male. So we started to get mentors and women who move forward in this career—and start to mentor them. Because I want to support them.

DB: What does your out outreach look like? Do people get sent to you? Do you actively look for folks? How do you find folks to enroll in the training program?

Brandon Smith: It's all and in between. What's that movie? *Everything Everywhere All at Once?*

Sadly, the state of California has thirty-five fire camps and training centers. And what a fire camp is . . . it's a half prison, half fire station, where currently incarcerated people work as firefighters. I was at one of them. We go to all thirty-five fire camps and say, "Hey, I know you may not think that you can go do this work once you come home, and you can. Because I've done it, he's done it, she's done it, we've done it. Look at this." So there's one inspirational point, and then they start writing us and we converse before they're released, and we try to help them out and refer them to services.

The next thing is, we have a six-month career training program, from November to April. We have people Monday through Thursday, we pay them $540 a week. They spend half of their time in class certifying as firefighters, and they spend the other half doing fire prevention work. We work with homeowners who have challenges with wildfires.

FFRP meets at a very, very unique intersection of needs. One is a climate conversation. The world is getting hotter and drier, especially in the western United States, which leads to more wildfires, and more severe wildfires.

Because of that, in the western United States, it's a criminal justice conversation as well. Because you don't have enough firefighters and forestry folks to go work in this space, you utilize currently incarcerated people to go do the work that you can't fulfill.

Yet when they come home, they can't continue in that work.

It's, like, "So you going to train me to go be a barber while I'm incarcerated, and yet I can't go be a barber once I come home?" Or you're going to train me to be a chef, but I can't be a chef once I come home? Yet you talking about rehabilitation? That's not a form of rehabilitation. Period.

There's also a conversation around [the fact that] the majority of people who operate in this space are white men from non-urban communities. I'm from LA. What you see in my back right now [mountains in Angeles National Forest], I saw this all the time, and I never knew that I could go out there and go do this work.

And so we go out there, and we go to the high schools, we go to the facilities. We're trying to stop people from getting incarcerated, and we want to say, "Hey, this job and career is available to you, and you can go make $80,000 a year graduating from high school. Come on, pull up, and we'll go train you."

My last thing is, when we talk about wildfires, there's a conversation around prevention and suppression. We operate in that space as well. I'm not here to just go train folks to be firefighters, but I'm also here to go train folks in prevention. Let's stop the fires before they happen. Let's go listen to what the Indigenous people said, and use their practices, and go out there and move forward and replant their vegetation, and use their practices so that these fires don't happen and get so bad.

One of the things that we think about as well is that today's firefighters are tomorrow's captains and chiefs and fire planners, safety officers, all that. And so we're trying to make sure everybody has as much information as they can so that we can all be great stewards of the land.

DB: Where do the trainings take place?

Brandon Smith: We have three training areas. One is in Pasadena, California. One of them is in San Bernardino, California. And we just opened our new office in West Oakland, in Oakland, California, in the San Francisco Bay Area.

After the training we help them to apply to different careers at Cal Fire, which is the state, the US Forest Service, which are federal agencies, and different private agencies. And come April they all end up being hired as wildland firefighters, and they go start their transition.

DB: Anything else changing over the coming years? What's on the agenda moving forward?

Brandon Smith: The agenda is to get more women. The agenda is to increase the way that we do fire prevention. We want to be available for more homeowners, fire safety councils, HOAs [homeowners' associations], all that kind of stuff. We just want to be land stewards. We just want to be here to go through fire prevention projects. So that and helping more folks.

Another thing . . . on the horizon for us is that we are expanding our career support to not just being a firefighter, and where you can be a forestry tech. For example, one of my participants was just working on a reforestation project. How do we talk about the water table and floods and all that kind of stuff? We are just here to go support.

There's so much need in this space that's not just firefighting, and that's where FFRP is expanding to.

DB: I'm curious if you got a surge of new training recruits or donations or anything after you were on *United Shades of America* with W. Kamau Bell. Do you think it helped change the perception of formerly incarcerated firefighters?

Brandon Smith: We were extremely blessed after that episode. After we filmed, Kamau posted us on social media, and because of that episode FFRP has raised over $45,000 just in PayPal donations or just online giving campaign. It has been a blessing. And I think the biggest thing that happened because of this was just connections.

One of the biggest challenges in my work is that people always ask me, they say, "Why? I don't want a firefighter to go steal my TV and save my house." First of all, when has anybody ever heard of a firefighter stealing the TV? Why is it that you just assume that us doing this work is going to be a challenge? We've been doing this work since 1941, since World War II. The state of California has utilized incarcerated people as firefighters in fire camps. This has been happening for a while. It's almost going on one hundred years. At one point in time, there were nine thousand people a year going through the fire camp system.

And there's this challenge where people think that, all right, I was formally incarcerated. Why? Well, let me say it like this. When I was locked up, I was like, "Damn, I fucked up." Like, "Damn, I'm a felon now. I'm one of them now."

Then when I was fighting the fires, and all these folks are waving and saying "hello" and "congratulations" and "thank you for all your support," I'm, like, they're not talking to me. My captain tells me, "Yes, they're talking to you. You just saved ten thousand people. You just saved their homes. These folks can go to school the next day. You are a public servant. Thank you for your work." I'm like, "Damn, I am a public servant." That's one of the things that drew me to this work.

And one of the challenges that people have is that, being what you would consider a public nuisance to a 180 change into being a public servant, folks can't put those two things together. It doesn't rack good in their minds. When you think of somebody who has been incarcerated or justice-impacted, you think of a drug dealer or somebody robbing your house or somebody that does something wrong. Then when you think about a firefighter, you think about the person that's going to get the cats out the trees.

DB: All heroic and white, yeah.

Brandon Smith: Right. You think about those folks. And they can't figure out how both of them come together. And it *is* possible.

I'm not saying it's possible in every situation, but it is possible. Because I'm living proof that it is possible. When you think about firefighters, people got problems with politicians, they got problems with teachers, they got problems with so many folks, like police officers. Nobody's got a problem with the firefighters.

DB: I'm glad that you brought that up. I think people want to compartmentalize, and they want to put people in their boxes, and they want to keep them there. It just makes people more comfortable to have predictable assumptions about what kind of character certain people have.

Brandon Smith: I've gotten lots of negative feedback from all sides of the panel. It's just hard, like you said, for folks to figure out where everybody stands in this conversation. And it's very interesting. I've had folks come to me and say, "Well, what makes you and I different is the fact that you got caught and I didn't. And you're from Los Angeles and I'm from Big Bear where the police ain't at."

So I received that as well. And then also there's this thing, that we all have the opportunity to serve and help folks out, and so that's what we do with it.

Marginalized communities, justice-impacted people, those who don't have their own seat at the table—all of us deserve the opportunity at gainful employment to support ourselves and our families. We all deserve the opportunity for . . . careers. All of us are the sum of all of our actions, and all of us together can be public servants, building together for all of our communities.

Listening to the Loved One
Wildfire, Community, and Ambiguous Loss
Amy Elizabeth Robinson

On September 26, 2020, I woke in the night, fumbled for my journal and pen, and scribbled down words that had emerged from the dark of dreaming.[1]

> What do you care about?
> In your vision, who is on your mountain?

The next morning, a spark of unknown origin lit a flame in Napa County, on the dry and rocky slopes east of St. Helena. "Oh, a fire," I sleepily said to my husband when I woke up, walking past him to a freshly made pot of coffee, trailing my fingers across his back. There had been so many fires that summer. Here was another fire. All that day the fire spread but stayed on those far eastern slopes. We lived across the broad, heavily irrigated valley, up the western slopes and over the ridge into Sonoma County from neighboring Napa County.

[1]. Written with gratitude to members of the Wappo community who have generously offered us knowledge and support over the past several years. We know we dwell on your land. Thanks also to Ursa Born for generative comments on an early draft and to the entire Monan's Rill community—present, past, human, nonhuman, and ghostly—for patience and companionship of many kinds.

So far away. But by 7:30 p.m. the fire had jumped and flames towered, visible from the road in front of our house, across our own small rugged canyon.

At about six the next morning, on September 28, 2020, our home burned. Despite the fact that we lived in an intentional community—a place where neighbors always knew when you were home or gone, when you had company, when you were in trouble; a place designed for connection and understanding—no one was there to see it burn. None of us were on the mountain.

· · · · ·

For some months before the Glass Fire ("our" fire), I was dreaming what a friend called "threshold dreams," dreams where the boundaries between inside and outside, human and wild, and damage and healing were all mixed up. As the shocking lightning fires of August 2020 turned the skies an apocalyptic orange in most of Northern California, I dreamt of a lush and fragrant otter cave clogged by human waste. The otters turned to me with pleading in their eyes. I dreamt of walking through the meadow at the top of our forested land, noticing parts of beloved oak trees burnt and dead. I reached out and touched a branch. It came off in my hand. I dreamt of strange and beautiful families on the land, while vultures circled in smoky skies. I had disorienting dreams of unfamiliar houses. In one I sat by a window and noticed a cemetery close by. I rose to open the door, and shadow creatures—moths and bats and vivid birds—flew frantically in and out.

"This is an edge time," said my friend. "An edge season."

I see now that there was a part of me working out my relationship to *place* and *wildness* and *shelter* in advance of disaster. But still, the loss, when it came, was unfathomable.

· · · · ·

Ten out of eleven of our community's houses burned. As did so many others in our rural watershed, all the way to the eastern edge of Santa Rosa. I am going to try to tell you about this loss. My sense of time, orientation, connection, and community itself have all been disrupted, wounded, refashioned, and transformed. It's hard to write in any conventional way when you feel like this. Today I am sitting in a newly built home, with the cat who traveled with us away from those towering flames sitting in my window, looking at the same meadow that existed before the fire. But the meadow is filled with a plenitude of rodents because the snake population has not recovered, which means that we also have new species of raptors soaring and diving above us now. The balance is precarious. But we see and feel some force restabilizing, in real time, in the forest around us, which is still filled with charred and bone-dry, utterly dead trees, as well as flourishing oaks and wildflowers. What I feel a lot of the time is wonder, and what I try to practice is a patient listening.

I am going to try to tell you about this. Because, though our community seems small and unusual, this is your world too. Precarity and disorientation and disaster, yes, but also ambiguity, liminality, flourishing, and wonder.

· · · · ·

When my husband and I moved into our house in 2009, with a two-year-old and another baby on the way, we stood at the railing of the back deck, smelled the thick firs, listened to the Steller's Jays' raucous *tschook-tschook-tschook*-ing. He said, "It's like our little cabin in the woods." I laughed and leaned into him. "It's not *like* that, hon," I said. "It *is* our little cabin in the woods."

Eleven years and four months and so much life later, I wrote this:

> My house is nothing more than emptiness, emptiness is
> nothing more than my house. Home is exactly empty,
> and emptiness is exactly home. My house is
> empty:
> Nothing is born, nothing dies,
> nothing increases and nothing decreases. My house is ash.

There is a photo of that cabin in the woods taken on March 8, 2020. I was on my way home from a hike, trying to walk out the anxiety about the global medical news that was encircling us, trying to decide if I should ask my family of origin to cancel an eightieth birthday celebration for my father. As I approached our quirky little house along the gravel road that stretched down from our mountain ridge, I stopped abruptly. I looked at the house and knew, with absolute certainty, for the first time ever, that my husband and I had made a good choice to live in an intentional community. To be in a place steeped in beauty and safety. To have neighbors (co-owners and friends) to call on for support and problem-solving and unexpected love and trips to the store.

I was, in fact, just emerging from a few years of what is usually called a midlife crisis but that I called "Everything I Touch Is a Question Mark." Since about autumn 2017, when the drastic fires of that year traumatized our entire county and came within about a mile of our home, I had been railing at the limitations of my life: consensus decision-making, conventional marriage, motherhood, the fact that I could not manage to fashion a "career" for myself. But in that springtime moment, as a pandemic descended around us, all that railing and pummeling and questioning dropped away. Things felt still and silent and steeped in a sense of enough. I pulled my phone out of my back pocket and took a photo. Now that photo hurts so much.

> Gone. Gone,
> gone over,
> so totally gone. My home
> is ash.

• • • • •

Our community, called Monan's Rill, sits (sat?) on a 414-acre swathe of the Mayacamas Mountains in northeastern Santa Rosa. The land, buildings, and infrastructure are collectively owned, but each family lives (lived?) in their own house. For three years, between the fire and a few months ago, several members lived in trailers on the land, one couple stayed in the house that did not burn, and my family rented in town. (My use of tenses and directional verbs became and have remained blurry over the course of these years. "Can you *come* up to our workday this weekend?" I would ask a friend, while even I had to *go* for thirty minutes to arrive. And just last night I sat talking with my daughter in my new bedroom and froze for a second, not knowing where I was and whether I had made it back home at all.)

At "the Rill," as we affectionately call our home, we use consensus to make decisions, believing that every voice holds a piece of wisdom and that it takes patience and deep listening to get there. We steward the forests and meadows and streams, knowing that we "own" them legally but that all land here is stolen by settlers. We *try* to honor and repair that history by being in relationship with the land, as best we can. And in the midst of this, over the past few years, we *tried* so hard to prepare for fire. We were informed, diligent, and motivated by a healthy dose of fear. After all, in 2015 we had learned that fire no longer followed the rules we had all taken for granted. That is when the Valley Fire rushed down from Cobb Mountain to the northeast, refusing to stay in forested land. It jumped the Lake County valleys and consumed unexpected towns, taking two thousand structures and four lives. "Sky's so dry you could light a match / by winking at the clouds," I wrote in a poem while that fire raged. I was embarrassed by how "their fire" made me so thankful for my own, for the golden big-leafed maples, for shelter, for the scent of later summer grass, for good sex and an intact family. A sense of wholeness. "There's a big hot hole in the land / up north and east that makes my / life feel glorious full, and all / my dreams feel edgy." An uneasiness had set in. A sense that some contract with something we called "nature"—a contract of which we had been blissfully, stupidly unaware—was broken.

And then, in October 2017, the Tubbs Fire roared across our mountain range just to the north, flew down into Santa Rosa proper, jumped the six lanes of Highway 101, and devoured the entire suburban neighborhood of Coffey Park. To our south on the same night, the Nuns Fire glowered and whipped through the southeastern towns of Sonoma County. Over thirty people died that night, terribly, tragically. Fire was not supposed to act like this. What was happening?

My community spent the next few years trying to learn, to understand, to get ready. We helped set up a watershed-wide emergency alert system. We invited the fire department to use our land for training so that they would know the roads, know where their helicopters could drop down into our meadows, know where the water sources were for refilling their tankers and hoses. We spent countless hours attaching fine metal mesh to the bottoms of our creaky wooden decks and over the tops of all the vents on our 1970s and 1980s houses. We thinned the fir trees that towered around us and filled the forest with undergrowth. We argued about what it meant to take down trees, how many was too many, what "safety" and "reason" and "wilderness" all meant in this new time. We did everything we could to prepare for this new kind of fire. But fire clearly wasn't watching. And when it arrived that night, late September 2020, the fire crews didn't. Human resources are so terribly thin in the face of a force so ferocious.

· · · · ·

Our house is gone. It can't come back. But so much is strangely left.

There is a forested ground to walk on, but it is a ground that will never be the same. There is a community to lean on, but it is a community that has changed shape and is painfully learning what it is and what it might become. It is a deeply confusing time. I struggled for months to find a way to explain the confusion of my grief, and then, in a conversation with my sister, I came across the term "ambiguous loss."[2]

My sister was talking to me about a family member who received a difficult diagnosis. She wondered if the term, coined by psychologist Pauline Boss, could help our family navigate the prolonged and uncertain progression of their illness. Ambiguous loss, she told

---

2. Pauline Boss, *Ambiguous Loss: Learning to Live with Unresolved Grief* (Cambridge, MA: Harvard University Press, 1999).

me, is any kind of loss that involves both presence and absence at the same time. I gasped. "Oh! That sounds like Monan's Rill. That sounds like us!"

Here is how Boss defines ambiguous loss: "a unique kind of loss that defies closure, in which the status of a loved one as 'there' or 'not there' remains indefinitely unclear. *One cannot tell if the loved one is dead or alive, dying or recovering, absent or present.*" These years since the Glass Fire have been a constant practice in living with that second sentence. But in our case the "loved one" of ambiguous loss has not been an individual or even our physical house and belongings—*Gone. / Gone, / gone over, / so totally gone*—but rather our community itself: the intertwined existence of people and land. People who loved and tended the land, people who loved and tended one another, the land that seemed to love us back.

At one and the same time, the land is torched and disorienting, and it is greening and regenerating. And though it is harder to see, I think this must be happening to Monan's Rill too. Many people have left. Rituals and structures have had to shift. We argue and cry about hard decisions we need to make. But some things are unshaken. We still value each other's voices. We still laugh. And what will emerge from this scorched reality is completely, utterly uncertain. Even now that we are back on the land, in actual houses, it sometimes feels like shadow creatures are flying frantically in and out.

· · · · ·

I have come to believe that every wildfire loss is an ambiguous loss. A person who has lost a home to wildfire walks the wounded earth where their house was as if in a stupor. The house becomes a ghost house, the landscape a ghost landscape (even when it starts to green again). The shell of identity that a person assembled over a lifetime becomes a ghost identity. Fragments of this ghost life appear nonsensically in the ash. A shard of pottery. A few lines from a page of a book. A warped, cast-iron bathtub. We invest so much in objects and routines. The turning of the corner in slanted morning light as your partner pours the coffee. The velvet box on the bedroom shelf that holds your grandmother's jewelry. The deep body knowledge that your children are sleeping in a particular room with particular comfort objects around them. The peep of towhees and the scuffle of deer in the leaves. Owlsong. Even the threat of scorpions. You feel the ache of all these absences like multiple missing limbs. You are alive, and the body of the land is resilient, yet nothing feels the same.

 I also believe that community wildfire loss is a particular shade of ambiguous loss, one that has surely been felt not only in my small community and our encompassing watershed since 2020 but also in whole neighborhoods and towns that have been consumed: Coffey Park and the Journey's End mobile home park in Santa Rosa, and also Lahaina, Blue River, Paradise, Greenville, Klamath River, Superior, Sagamore, and more that I cannot name. My heart can barely hold them all and our loss too. These are the new ghost communities and ghost towns. In each of these places there is this larger being that has been devastated and unwillingly transformed. The paths and living connections that wind and stretch between homes, not just the separate homes themselves, are covered in ash, hard to reconstruct. And in each place there is the stretch of time in which confusion, ambivalence, immobility, waiting and wondering, good-byes, startling reappearances, creativity, numbing (or enraging) bureaucracy, astonishment, hurt, absurdity, and resilience all co-exist. A stretch of time that has no clear ending. A threshold.

 Like others who have struggled with intense loss, I questioned how to best cross that threshold and whether to return to my community. I had months and months of wandering, disorienting dreams, familiar paths and places opening into unexpected landscapes or destinations that are just plain wrong. My consciousness was always wondering what path to take and where to land. And every once in a while I would have a deeply pleasant dream of some strange, new house. Sitting down to a good meal in good company. Lush, dark forested artwork on the walls. Once I chose to move into a mansion, huge and decrepit and cluttered, full of eccentric family members who could tell me its stories. It was in a bustling urban area, on a waterway. *I loved the place*, I wrote in my journal. *It felt alive. The house felt dead, but it seemed like together we could make it, too, come alive. It would be a long commitment. But I'd be close to family, and we'd have a sense of purpose—caretaking, not just sliding into normalcy.*

· · · · ·

When almost everything burned on that Monday morning, nineteen adults and eight children were living at the Rill, and a new family was just about to move into an open house. That new family had been at the Rill on the Saturday just past, for a community workday. First the mother and I sat facing each other on a rough bench to sign the rental contract, giddy with the potential of this new life together, and then we worked to rake dry oak leaves away from the edges of our community building called the Hub. While we raked, her child and a few community kids drove tricycles madly (masked and slightly distanced) around the concrete game court behind the Hub, getting to know each other. At lunchtime our families sat on picnic blankets about six feet apart by the swing set across from our house and talked about chicken coops and outdoor distance learning and feminist philosophy.

Everything felt so possible and vibrant.

When almost everything burned two days later, we were all scattered, into friends' houses and hotel rooms across Sonoma County. Evacuated. Waiting.

Two days after that, all the grown-ups gathered in little digital Zoom boxes to hear the strange and dreadful news. The Hub was still there, our community workshop was still there, our tractor and truck were still there, parked in an open grassy space across from our orchard and garden. One single carport and one single woodshed were strangely still there. Just one house, at the far western edge of our residential area, was still there. The tricycles left sprawled on the concrete and the brightly-colored chalk drawings that the kids had made on that Saturday: still there. Two swing sets, one in the weed-whipped grass across the road from our house and a matching one in the back of the community garden: still there. Everything else—ten homes, plus our community toy shed, garden sheds, greenhouse, chicken coop and many beloved chickens, well house, orchard trees, fences, barn, barn cat, my writing studio—gone. Black. Gray. Ash-white. Dusty brown.

Incinerated. Gone.

"Some aspects of the [loved one] are lost forever," Boss writes, "while others are very much present. The task is for families [or communities] to remain aware of the difference." To clarify who and what—which people, what things, what qualities of being—are still with you or just plain gone. Boss is insistent, though, that the loss itself "cannot be clarified." This lack of sense makes sense to me now.

People say that wildfire burns in a mosaic way that defies logic, no matter how hard we prepare or try to understand. It amplifies the confusion. Still, I find even the mosaic metaphor to be too flat, too shiny. This feels more true: *The fire came like a terrible, moody dragon across the mountains from the east, devouring some things and leaving others. And it is still laughing.*

· · · · ·

My family lived in our rental house for almost exactly three years, surrounded by irrigated lawns and two-car garages. Through those years, I struggled to find the word to use when describing my fellow community members who were trying to rebuild together. My neighbors? Not anymore, and that doesn't capture it anyway. My friends? Not enough. My partners? Well, yes, in a business sense, but it sounds both too intimate and too formal. My fellow community members? Too bulky. My family of choice? True, but to me it somehow diminishes my family of origin, who remain crucial and dear to me. My Rillers?

There are nine of us grown-up Rillers left, plus just my two children, now thirteen and sixteen. The scattering of the "loved one" that was the Monan's Rill community began immediately and was confusing and prolonged. The family who was about to move in, of course, did not. Another family piled in a car and drove to the East Coast and did not come back. A third family was already going through hard times, and the partner who wanted to return has been struggling with illness. Their children have only been back to the land a handful of times, to ride those tricycles and see the pit that was their house. One elderly Riller who had already been transitioning away from membership also never returned.

And five others, who expected to leave in the next two to ten years (maybe?), came to meetings and workdays for a while but—two, two, and one—ended their membership and shifted their attention to town.

One last family tried hard to stay connected, but the pressures of young children and economic life, as well as the deep uncertainty around rebuilding timelines, took over, and they bought a house elsewhere.

Some people come back to the Rill now and then, which brings solace and joy but also sadness and confusion at times for everyone. Every "we" question became fraught after the fire. Who would be on the community email lists? Who would get a say in forest stewardship decisions? Who would be along during the raw rituals of grieving? And when we learned that the complexities of collective ownership, under-insurance, and the skyrocketing costs of construction tangled together to prevent us from being able to rebuild everything we lost, it became harder to envision a future we too. What was the Rill anymore? What were we to become? I now understand the dread and shadow that creeps across Emily Dickinson's poem "A Light Exists in Spring":

> *A quality of loss Affecting our*
> *content*
> *As trade has suddenly encroached upon a*
> *sacrament.*

Every community moving through wildfire loss is a community being built from the ground up: from the painful shadows and hard facts of what was lost, the strangeness and resilience of who and what remains, and the murky promise of what might come. Every story of community wildfire loss—whether the story is set in a rural place or a suburban place, whether the community was apparently close-knit or seemingly walled-off and atomistic, whether it happened to people whose homes were clearly at risk or to people who were stunned by the arrival of fire—includes this prolonged sense of disorientation, an extended grief of not knowing who and what to say goodbye to, or when, or how. It's almost as if we don't know the shape and beauty of our communities until they are gone.

· · · · ·

Pauline Boss tells a story about a woman who experienced ambiguous loss when her pilot husband was downed and his body never recovered. The woman, who grew up in an Indigenous community, told Boss that her husband had returned twice after his disappearance to speak with her, helping her make decisions about moving forward. "His symbolic presence provided direction," Boss writes, "[and the woman's] story forever changed the way I think about and do research." In another place in the book, Boss talks about the importance of overcoming the solitude of such confusing loss through honest conversation. In this conversation, the loved one's voice plays an important role, even when some people may feel that voice is too diminished or damaged or ghostly. It is important to offer the loved one respect, to hear and heed their words. It can help you move toward understanding in a time of grief and chaos, scattering and darkness.

When I lived at the Rill and felt lonely or isolated from life "in town," walking the mountain always soothed me, surprised me, and reminded me that companions surrounded me all of the time: jackrabbits and brilliant fuchsia, silky maroon manzanita and bright-eyed deer, and always the squirrels and woodpeckers busy in the oaks. So much life! So much company, if I could listen! "Be still and listen," says Wendell Berry,

> to the voices that belong
> to the streambanks and the trees and the open fields.

> There are songs and sayings that belong to this place,
> by which it speaks for itself and no other.[3]

Through the past few years of fire and recovery, the land has not ceased speaking to me, twining through my dreams and my waking life. I heard the forests sighing and releasing during that long, apocalyptic 2020 summer. I heard them say, *We've been waiting for this*. When my house was still safe, my children tucked into their beds with all their familiar things around them, I could step aside from anthropocentrism and see that other species, other voices, were asking for—and taking—what they needed. The forests were too dense after two centuries of violent, un-listening colonialism, and they needed this.

When Wappo cultural leader and fire ecologist Clint McKay walked the land with us after our fire, he looked out over the ravaged valley and said, "You have been given the gift of visibility." I try to remember this. I close my eyes and hear my son's voice as we drive up our long gravel drive, on the day we first returned to the land: *You can see the shape of the hills now!* I try to remember him running the deer trails that became more clear in the scorched meadow above our incinerated house. There *are* times when all I can see are the blackened manzanitas, the bodies of toppled firs, the gaping wounds of so many lost houses. When all I can feel is the fatigue of rebuilding. But then I close my eyes and listen.

We are relearning how to be in relationship with the land and with one another. We are reaching out for guidance and struggling honestly, the nine of us who are left. Both the land and the community are wounded and confused but also filled with a quiet strength that has been surprising. On a winter day in early 2021 we headed up a hillside meadow that had been compacted by utility trucks that had taken out precarious trees right after the fire. We carried buckets of native grass and poppy and lupine seed, and as we scattered the

---

3. From Wendell Berry, "2007, VI ['It is hard to have hope']," in *This Day: Collected and New Sabbath Poems* (Berkeley: Counterpoint, 2014), 305–6.

seed we also scattered hope and belonging across the ground. As the sun rose higher and my body grew tired, I lay down on the sprouting earth, looked toward the sheltering mountain, and heard a warm and plural voice as clear as day: *Thank you for not abandoning us.*

· · · · ·

Among my many threshold dreams during that long and devastating summer was one that offered more than confusion. It offered a vision that is still unfolding. It was the one with strange and beautiful people on the land. For those people were in a place that felt clear and joyful and full of song. There were children playing, young women bathing and singing, families pouring down a hillside from a ridge that was in waking life still the site of two houses (and which is now the site of two dusty bowls that once were houses) but that in the dream was a new gathering place. My husband and I drove through the land in our community truck. I stood tall in the bed with a clear view of the skies.

To the east the air was thick and smoky, but we could see one majestic mountain peak vividly. To the north there was rain descending from a cloud of ash. And from the west, from a range of dark clouds and trees, those vultures emerged: black, raggedy, flying in formation, rolling together, smoothly shifting places, seemingly dancing. "I didn't know vultures could do that!" I exclaimed, before we turned the truck and headed back to the group.

I see now that we are the vultures dancing. We are the cloud of ash, and the solid mountain. We are the community coming into being, again and again.

I continue to have threshold dreams, but over time they have become more about rebuilding the Rill than feeling lost. In one, there were so many people on the land! Workers and their families camped out, children on blankets just beyond the warm light of cooking fires, humans caring for each other. Folk dancers practiced their art, and I helped them with their music and timing. The Rill felt like a small, abandoned village, with rickety shops struggling to stay open and keep their business lively after the fire. But still dustily, doggedly there. Our home was on top of a bank, all burned up but being reframed. I wondered, in the dream, if another family would one day live there.

And now we are in a house at the Rill again, a long green house with a broad patio and a view of the garden and mountains. This is the choice we made, that we were *able* to make, and the beautiful, arduous journey we have been on. The last months of rebuilding were filled with details of the material world: baseboard and faucets and vanities and paint colors. Concrete and drywall and asphalt sometimes felt like they were squeezing all the creativity and connection with nature out of my body.

But we made it. We open our eyes again to oaks and deer and the raptors soaring. And after a few months of nesting, I finally went for a long solo walk again on the land. At the far northern edge of this place where we dwell, there is a small patch of woods, bordered on three sides by dirt and duff-covered roads, that miraculously did not burn. I had walked a few times past this patch of unburned trees, but I had never thought to step *inside of it*. So, as I stepped from a forest of logged firs and burned and stump-sprouting madrones into this small space of what-our-home-used-to-look-like, I gasped. A thick, cushioned leaf layer covered the ground. Vibrant fern-like moss draped across the tree trunks (*fairy forests!* my children and I used to say). The madrone bark was the burnished, peeling red that it is "supposed" to be. I reached out and touched its soft inner bark in genuine wonder. I felt like I was standing inside of magic, inside of *presence* instead of *absence*. The two are always so close, separated by clear air and the whims of life's dragons. Just one step and we can cross over. All of this possibility lives on this mountain.

"The world / is no better than its places," says Wendell Berry. "Its places at last / are no better than their people while their people / continue in them."[4] To continue in place can be strenuous and is indeed not always possible. Compassion for the difficult movements people choose or are forced to make would go some way to ease our

---

4. Berry, "2007, VI ['It is hard to have hope']."

human arguments and pain. This planet is on fire in so many ways. But when it can happen, to continue in place—listening, witnessing, walking the ground that is stubbornly greening, revealing flowers and grasses and ways of being that have been patient in their seeds longer than your lifetime—can be uncanny, magical, and healing. So I ask you now, as my dreams during these long, ambiguous years have been asking me:

What do you care about?
In your vision, who is on your mountain?

Parenting in Fire Country
An Interview with Kailea Loften
Dani Burlison

Kailea Loften is a Black American, citizen of Liard First Nation, and member of the Tsesk'ye clan from Nalokoteen (end of the ridge nation) of the Tahltan Nation. For the last decade she has worked at the intersections of climate justice, spiritual ecology, and independent publishing. Since 2019 she has been the coeditor of Loam. She has guided climate change policy with an emphasis on Indigenous rights on local, national, and international levels, previously serving as a climate commissioner for the City of Petaluma, and as the climate justice organizer and community publisher for NDN Collective.

In 2018 she published an essay at loamlove.com titled "Mothering into the Anthropocene." In it, she wrote:

> 48 hours before the Tubbs fire ignited, I gazed down at a positive pregnancy test in London. It was the tail end of a two-and-a-half-week trip to Europe, where I had been facilitating my course, "Earth Is `Ohana," a class on embracing spiritual ecology as a response to our climate crisis. The day before I took this pregnancy test I was sharing about the "3-6-9 concept," which references that our planet is headed toward a 3-degree Celsius global temperature increase, while going through the 6th mass extinction of species, and

simultaneously headed towards a global population of 9 billion people by 2050. I had asked the class to sit in silence and reflect on how the sum of these numbers impacts them. For the majority of my twenties, my identity has largely been wrapped up in youth and climate work; understanding and sharing information like this has been the focus of my life for the past few years. The enormity of discovering that I was pregnant was also intertwined in the reality of these numbers. I was excited, yet critically aware of what it meant to bring another human being onto this planet.

The essay struck a chord with many parents and nonparents alike across Northern California's fire country, and it sparked conversations about what it means to raise the next generation in a world of recurring climate disasters. In this interview, Kailea shares where she is now with fire preparation, parenting, resilience, and more.

DB: For people who maybe haven't read your "Mothering into the Anthropocene" essay, can you give an overview of what your focus is?

Kailea Loften: "Mothering into the Anthropocene" was the essay that I ended up writing after I had my son in 2018. He was born in June, and I think less than a month of him being earth side we already were dealing with fire and smoke.

We were all shuttered into our homes, and I was still very new to Petaluma and didn't know anyone. I was experiencing a lot of isolation; postpartum tends to be isolating, but I just remember these days where I could not take my son out for a walk because the smoke was so intense. I ended up going—it was actually our very first overnight trip when he was a little over a month old—to go occupy the front of then-governor Jerry Brown's office at the state capitol in Sacramento with a group of people that were organized by Sunrise Movement. That was my son's very first direct action. I remember feeling horror while walking into the capitol because the smoke was so bad. I had a blanket over my son's head, and I don't think I had a proper mask or anything. I remember how weak and exhausted my body felt by the heat and smoke exposure.

"Mothering into the Anthropocene" was written from that place. I was having this experience of being, within my organizer friends, the first person to have a child and feeling like my pregnancy was initiating everyone. The essay speaks to this transition from being a youth organizer to becoming a parent for the first time in a moment that is marked by climate change.

I found out that I was pregnant forty-eight hours before the Tubbs Fire broke out, which took place twenty minutes from my home. The foundation of my entire journey into parenting was anxiety and uncertainty and fear around what could happen in the night, what could happen on a windy day. [The essay] was meant to provide a bit of an inkling for a lot of my friends and for readers at Loam who were not yet at that place in life. There's going to be more of us young climate activists stepping into parenting, and I wanted to provide a really honest reflection of my experience.

Over the years people continue to reach out to me about this piece. Some are climate activists, some are other parents or people who are caretakers, some are just regular people. It continues to surprise me that the piece keeps circulating because when I wrote it I was really hazy. I was learning how to nurse a child. I was learning how to put diapers on. Initially I felt critical about it, thinking *This is not even a very good piece of writing, but here it is.*

DB: I think it's beautifully written. My kids are older, and my oldest kid's first direct action was at six months old at Headwaters Forest, trying to prevent old-growth redwoods from being cut. It impacts us, but not in the same way as a fire, because we can leave and come back home and feel safe and protected in our homes. And I think that what stood out to me so much about what you wrote is that sense of vulnerability that comes with living in this fire ecology, this fire country.

And as a new parent you're already so vulnerable. So, bringing both of those pieces in, I think it's a perspective that's really needed. People are mostly worried about houses and worried about getting out. But then mixing that level of vulnerability in with the vulnerability of a new parent, that's why I loved it.

I'm curious about how you're feeling now. Your child is older. You've been in the area a little bit longer. We've had other fires. There are other things going on, like the pandemic. I'm curious about where you're at emotionally and spiritually with parenting during these crazy, terrible fire times.

Kailea Loften: It's hard. If not for a severely traumatic moment in my life as an adolescent, I would say this would be the hardest time in my life. I had these other excruciatingly painful few years when I was coming of age where I had less perspective; I had less control over my own life. I had less emotional tools. So I would say this is the second-hardest time of my life. There are just so many things that I wish I was wrong about.

I'm very practiced in watching and listening in between what's being said, and it almost gives a quality, a feeling, like I can see into the future. A lot of what I foresaw in my earlier twenties and even at the beginning of my parenting, that first Tubbs Fire for me brought this moment of knowing it arrived.

It often feels like I'm betting against myself all the time on this deep mental level, because I wish I was wrong about what we're experiencing and what we're going to experience.

And I'm not. We're not. The people that I have worked with have not been wrong and have been very on point. During the summer of 2021, there were those huge floods that happened globally, and the heat domes across the Pacific Northwest. I went into a state of intense anxiety because my projections were behind what it is that we were seeing. I knew those things were coming. I just thought we had about ten more years.

I put a lot of emphasis on building and trying to maintain my backend systems, which looks like having a counselor and relationships with people where we can tell the truth with each other.

I think I know enough about myself now to know that I, as an organizer . . . can be in conversation with people who don't understand, but on a very personal level, my intimate circle of people, I cannot. I have to have people who really are seeing and understanding and listening on the same level that I am.

DB: That's so important. Because I had kids when I was very, very young, a lot of my friends' kids are younger. Most are school-age kids, and some of the parents really want to have those conversations as well. And then some of them can't even go there. And then when there's a fire, they are frozen with fear, and they don't know how to talk to their kids because they don't want their kids to feel the fear.

I'm curious if you have advice for other parents that are in this area, in this fire ecology, and that maybe don't know how to deal with it, don't know how to talk to their kids, don't know how to process what's happening as parents.

Kailea Loften: I think it's important to say that you can be an optimist and a realist at the same time. One of the things I've noticed is this weird thing with our culture where we're not allowed to really be the bearers of bad news. Or we're stuck in a toxic positivity loop.

We are in compounding crises. The climate crisis is the most intersectional of them all. I'm biased. I think we need to focus on it. I feel like if we all focused on it, we would be able to tend to all of the other issues, but I understand that people are bogged down by also being stuck in capitalism. The grind right now of having to maintain a home, a life, especially in California's economy, is really hard.

I hear feedback from parents [who say] "I need to shut that off because it's too much." I really understand. But I also hear from people, and I've heard this in several different ways: "I'm an optimist." There's this idea that if they just keep hoping or keep wishing or something, it'll be okay.

It's not about not being an optimist. I think it's a coping mechanism on some level, and I think it's an important thing to be protective over your visioning space and to continue to hold your imagination as a zone of sacredness. Because if everything eats away at that, then we're all lost. Right?

DB: Yes, absolutely.

Kailea Loften: But I do think it's important to be honest about what it is that we're experiencing and going to be living through, and what our kids are going to be living through, and what our grandkids are going to be living through. I don't think that it has to take away from our visioning space. I think there's space for both in us.

I don't know how to say it any other way other than we actually need to tell the truth about what it is that we've experienced. We have to continue to build collective memory. I think people have amnesia because of the trauma of what we have been experiencing all these years now. And we have to continue to remind each other because, if we don't, then we're not adapting and we're not preparing.

It's less advice and more just the message that the world is different than what you grew up with and the world is different than what you were promised. We all need to come to terms with that. It's not what you wish it to be, and we have to step away from that space of wanting to go back to normal. Especially because that normal was what led us here.

DB: I really appreciate the perspective too with being overwhelmed, not just with climate change but also with capitalism. And I keep thinking about how I feel like most of us haven't had time to really slow down and process these catastrophes as they're happening, because we're expected to just act like they're not happening. And we're expected to keep consuming and producing and working and working. What I'm seeing from personal experience is that there isn't a lot of space here for us to really process. So, if we're not able to process, I think it's harder for us to help our kids process it. It's kind of this fast-moving hamster wheel that we're on.

Kailea Loften: I think too, with kids, one of the things that we really advocated for on the Petaluma Climate Commission is bringing in education throughout our schools on what climate change is and specifically what it means to live here.

One thing I want to say to parents is that it is important to start conversations with your kids on these topics, for a few reasons. One is safety, which is critical. If you're not starting conversations with your family about evacuation and what is expected, then you're actually not keeping your family safe. And that's something we all have to be in a conversation about for those of us that live in California.

I have run a lot of scenarios with my husband, and every year our scenarios shift, depending on how heavy my kid is. Now we're at a point where he's old enough to carry his own backpack. So, if we had to evacuate, heaven forbid, this year, we would give him his own pack that he would have to be responsible for. Safety is a crucial part of all of this. It's a part of how we make sure that we're tending to each other and our wider community.

I am upset that our own parents did not start conversations with so many of us, even though the writing has been on the wall in terms of so much of the systems' breakdown we're experiencing environmentally, politically, socially. Many of our parents stayed in denial and are still in denial.

I remember when Wall Street crashed, I was sixteen or seventeen, and it was intense. It tanked our economy in Hawaii, and at the time my mom was a massage therapist. I remember her going to work, sometimes only making thirty dollars. It was awful. It stripped away so much for our family. Yet my mom still wanted me to continue with this trajectory she had mapped out for me with my college plan at the time. With what limited information I had about the world at that moment in time, I remember thinking, "What she has mapped out for me does not match the reality we are in."

I took it upon myself to make some big life choices then, that I'm still grateful for but at the time were not affirmed by any adult around me. No one could understand why I suddenly withdrew from school. Now, in hindsight, people can understand why I didn't want to take out a lot of student loans, which would have mired me in debt during a financially unstable time. But that's the thing: when we don't actually take the time to really look up and see what's happening around us, and we're not honest, and we don't start those conversations with our kids, we simply do not prepare them for what we're going to live into, and why would we do that to our children?

DB: I agree.

Kailea Loften: In terms of this advocacy to bring more education into our schools, one thing I want to say to parents too is that you shouldn't have to carry it alone. This should not be a conversation that exists solely in your household. This should be something that our community is holding.

I really want to say "I see you." How are you supposed to hold every conversation of precedence in your household right now? That's not fair. There are things that need to be said in our homes specifically around safety, but there are also other parts of the conversation that need to be permeating our whole social sphere, because it's too much for individual households to hold alone for their children. That's not okay either.

DB: From a parental perspective as well, I'm talking to a lot of people who are not only afraid for their kids' future and our collective future and having a hard time talking with their kids but that are also having a spiritual crisis, honestly. A lot of folks are having a crisis because of what's happening, trying to be realistic, trying to maintain this connection with the planet while also kind of being afraid of the planet because of these climate catastrophes we keep having.

I'm wondering if you could speak to that, share your perspective. I know there's no solid answer, but how can people maybe find balance between recovering from the spiritual crisis, or how to move forward with what we're facing.

Kailea Loften: That's a great question. I initially ended up in Sonoma County because of my partner. He's a documentary filmmaker, and I met him on a trip that I was part of for this spiritual ecology youth fellowship.

This fellowship was all based around the field of spiritual ecology, which is the recognition that we are in a deep spiritual crisis with ourselves and ultimately with the earth, and a big part of this space of exploration is this understanding that a part of what has created the crises is a break in conversation with ourselves. In order to actually address the large-scale crises that we're experiencing and living through, we actually have to have a very alive and robust internal conversation with ourselves first.

The word that's popping up for me is "reciprocity." So, for those of us that take a lot from being in connection or in relationship with natural spaces, I feel like spiritual ecology is asking us to reciprocate by being in a place internally where we understand how our own actions and our own lifestyle choices and our actual ways of living encroach on or impact the natural spaces that we also take so much from.

One of the things I love about this field is the understanding that we really just need a values-based way of living again. And that is what so much of religion, faith-based spaces, and spiritual practices provide us, across the board . . . values to return to.

I know that there's a lot of people, especially in a place like Northern California, who don't necessarily consider themselves religious or a part of an official church for varying reasons, which I respect. But because of this big departure that has existed within our society of leaving churches, leaving religious spaces, there is a void that has been created. We see that people are spiritually hungry. We see that people take, which is part of why appropriation happens.

There's this sensation of having a hole in oneself where there should be something. I think we're looking for . . . values and accountability, to actually live out and practice and something to come back to when we're struggling or when we're just not living in alignment with our own personal philosophies.

That's my assessment in terms of that breaking point. I guess what I would say to people who are feeling that place of fragmentation is twofold. One is that the grief is real. It's really hard to love something or someone so much that you are unapologetic about it.

What a vulnerable and beautiful thing to love loudly in that way. And also it's going to hurt when you see that thing or that place or that person be harmed or be disappeared in some way.

I think it's important for people to do that grief work, find the communities who can hold and are presencing with grief at this moment. There are so many, and there will continue to be more. It's a practice, and it's important to do because it's energy that needs to be moved. And we should grieve because grieving is part of how we acknowledge.

We need guideposts as we navigate what feels like a lot of darkness.

DB: Yes, I think grief work is so important and especially in this context, because it's not like things are going to change for the better overnight. I agree that acknowledgement and that processing is hugely needed as well.

I don't know if it was the last fire or the fire before that, but I think it was Sonoma County supervisor Linda Hopkins who was quoted as saying something like: "We're all a little resilienced-out at the moment." That term is something that stuck with me: "resilienced-out."

I know it's important for us to have resilience in these times, but I'm curious about the things that you do personally to help yourself with that sense of resilience to keep moving forward.

Kailea Loften: Such an important conversation in terms of resilience right now, and there's layers to it, especially within Indigenous communities, within Black communities. There's this sentiment: stop calling us resilient because to call us resilient is to, in many ways, invisibilize actually how traumatized and harmed we are.

We're tired, and in this context we should stop throwing this word around in such a casual way.

On another level, resilience is not really fully understood. And that's why I say "thrown around in a casual way." There is a place that is called the Stockholm Resilience Center based out of

Sweden that I did a course through some years ago now called "The Planetary Boundaries and Resilience Thinking." The Stockholm Resilience Center studies and maps out what constitutes a resilient system and has identified categories within systems that have to be built up, like a muscle exercised and practiced in order for a system to be considered resilient. So we can say on a very surface level that this person is more resilient than that person, or this community is more resilient than that community, without really having an understanding of why.

One of the things that I love about the study of resilience is that at the core it's about a system, a community, or a person's ability to withstand shock or crises, and also to be able to come back after a shock or crisis. Again, I totally respect and understand this sentiment of "we don't want to be called resilient anymore." But when I think about city planning and I think about adaptation, we actually have to work with resilience as a framework because the reality is, if we do not, it means that we're not adapting. Rebecca Solnit in her book *A Paradise Built in Hell* goes and interviews communities after crises. The thesis and the sentiment that comes forward from people is that during the actual moment of crisis, people felt there was this out-of-time moment where there was some form of a "paradise" built. A place where people dropped political ideology and barriers and found humanity, found each other.

I think it's important to also recognize that one aspect of resilience is this understanding that there's opportunity, always, that exists within a shock or crisis. And we know that based on Naomi Klein's work *The Shock Doctrine*—that companies and other vested interests have always seen and understood that there's opportunity when there's a shock or crisis. But if our communities and our people don't understand that, then those vested interests will continue to win. We have to become a lot more practiced in seeing the opportunity within a disruption in the system so that we can actually bring our values and use those moments to start building, practicing, codifying even, what it is that we want for our communities, because if we don't others will.

We, of course, need time and space on an individual level to be broken down, to be held, to rest, and to not have to contemplate the next step. But we also can't stay there. That is not a privilege that is allowed to anyone right now. And those that think that it is are in denial, because we're going to keep experiencing more shocks and disruptions stacked on top of each other.

DB: Thinking about this idea that individuals need to be resilient and have this resiliency . . . but that actually the resilience comes from having a community to support you and help you through these hard times and to really see you and listen to you and reflect back and all of those things. . . . Thank you so much for all of this. Do you have any last thoughts to share?

Kailea Loften: I think that in terms of a sum-up, I would add that something I found that people feel very harmed by right now and feeds, I think, into this space of toxic positivity is that we're being constantly gaslit. I have heard from a lot of other mothers that reading my essay or simply just exchanging stories on what we're living through has been affirming. People have told me "This is the first time that I have been told I'm not overreacting. This is the first time that I feel empowered to take the next step, which is to prep, to make my plans. I've been told that what I'm experiencing or what we're experiencing is temporary or not that bad, not that extreme."

And then at large, for all the readers out there, it's important to name that we're all being gaslit by politicians, the fossil fuel industry, and capitalism at large.

We continue to be fed this narrative that everything will go back to normal and that we can continue living as we always have. I think we're all selling ourselves short when we believe this narrative. It's comfortable, and it's no longer grounded in reality. We're experiencing collapse alongside growing pains. I want to live to see what's on the other end of who and what we're all growing into.

Funding and Finances During and After Disaster
Sue Weber

With the exception of war, most disasters, if not all, happen in a moment. Tornadoes, hurricanes, floods, and earthquakes usually happen within a very specific time. The disaster strikes, and the people affected are left to deal with the aftermath of destruction a few days later. This isn't true for wildfires.

In the case of the largest single non-complex wildfire in California history, the Dixie Fire, where nearly one million acres of forest and communities burned, residents could not immediately come back, assess, and figure out how they were going to move on with their lives in a timely manner of a few days. The evacuation of communities in the Dixie Fire began in mid-July 2021 and lasted until the end of August for some affected communities. A six-week evacuation is a really long time.

There is a substantial financial journey in the picking up of the pieces after a fire. It is a microcosm of what happens at larger scales. Financially, people had to have had some type of savings in order to address their needs. But we found that not to be the case for many of our residents. Greenville households before the fire had a median annual median income in the $30,000–$40,000 range owing to the large number of retirees and low-income residents. People had to find safe housing and be able to access food and water. Some, since it was summer, found a spot on the river and camped. A few found their way to Red Cross shelters. Those who could afford it stayed in hotels or with family members farther away.

In Plumas County, some people bounced around for well over a year after the fire, some living off their credit cards or on prepaid cards that were donated. People camped in their friends' and family's yards and properties. Some parked FEMA trailers on designated lots until there wasn't financial assistance to do so anymore.

If you had insurance there was almost immediate relief. Insurance companies, like State Farm, had money in folks' accounts by the next day or a week after the fire. Those who were not insured—well over half our population—were left fending for themselves.

Finding any type of open housing was nearly impossible in Plumas County before the fire. After the fire it only got worse. The town was lost for everyone. Even for those few whose homes were not lost, there were no services. There was no gas station for months. The post office finally reopened in August 2024.

In order for a community to receive FEMA funds, the president needs to declare a state of emergency. For bigger communities this happens quickly, as more people are affected and more property of value is destroyed. In rural areas residents are left to wait for relief until a threshold number has been reached of lost structures or their cumulative value. Since forests burning do not count in this equation and houses valued at less than $250,000 are under that threshold, communities are in a holding pattern for relief. This was the case after Greenville burned. The federal government seemed to be waiting to tie in multiple fires in what seemed to be a way for them to justify the numbers. But there was no immediate relief.

During this time organizations such as the Red Cross and United Way set up sites in conjunction with a few local nonprofits to distribute immediate and effective services for those impacted. There were gas cards. Gift cards to nearby stores. World Central Kitchen came in and helped with meals along with local folks stepping up and providing for their neighbors. Wealthier middle-class residents put up in hotels in Reno, Nevada, two hours away. Others parked their fifth wheel RVs at friends' places in neighboring towns.

All this took place while the Dixie Fire still raged and with folks in the non-impacted areas on edge, wondering if their towns would be next. How long can people live this way, with little sleep and little finances? Some employers helped their workers and recognized what they were going through. Some employers went on, business as usual, acknowledging very little about their workers' predicaments and changed needs—such as time off when they ran out of vacation time to deal with what the fire had brought them.

Once the Dixie Fire was declared a national disaster, FEMA funding was initiated and began to trickle in, but the challenges were daunting for many needing aid. Adult children whose single-wide trailers were parked on their parents' property had to struggle to provide proof that they had been residents, even if they'd been there for decades—a common enough practice in rural areas. Even the most educated person could have trouble navigating the paperwork involved with FEMA and making any sort of claim. Indian Valley prior to the fire, according to the last census, has the highest number of PhDs in the county but also the highest number of high school dropouts.

My observation in a nutshell: those who had good insurance did not qualify for or need FEMA funding. Those who were poor and marginalized usually received the maximum funding. Middle-class residents received very little if any funding and would become the population that would struggle the most in rebuilding.

With so few nonprofits in the county, getting monies to flow in was challenging. Most local nonprofits did not have the capacity to accept large amounts of money. Organizations popped up claiming they were raising money for victims of the Dixie Fire, and locals never saw any money from them.

Almost immediately after the fire, but well before anyone was allowed back, the long, arduous process of cleanup began. The fire burned so hot that some folks came back to empty lots without so much as a few recognizable bathroom pipes left. Folks who had homes to come back to returned to smoke-filled homes needing specialized cleaning, perhaps slight damages, and refrigerators filled with spoiled food. They returned to unlivable spaces. They returned with finances stretched thin and struggled to cover new recovery expenses at the beginning of a new school year.

At the same time, the unprecedented disaster left an uncreative and un-generative county government even more paralyzed and dysfunctional than before, with little leadership in how to move forward. With limited experience and skill in governing, the board of supervisors wanted donated money to go to the county's general fund rather than to specific fire-hit areas. Some supervisors chose not to visit these affected areas for months because they felt it would be too depressing. It could be all too easy for them to ignore the problems of recovery if they did not travel through or live in the affected areas. More accountability from elected officials in rural counties is needed moving forward. Residents must demand more from them. Some supervisors who were in office in 2021 in Plumas County and had held office for years have already been replaced in elections since then.

As nonprofit organizations and other entities, like community-based Plumas Bank, began to raise funds for the disaster under the guidance of North Valley Community Foundation, which had experience working on the Camp Fire three years before, the Dixie Fire Collaborative was formed. A Funders Roundtable was created, which was one of the most brilliant things to come out of the fire experience. Organizations began working together by pooling their dollars and collectively started meeting the needs of individuals at a grassroots level in less time than government agencies could deliver. Weekly and then monthly community meetings took place to have as much local-resident input as possible in determining what the needs were and how to allocate funds. This began well before FEMA kicked in. The state, it should be said, almost the day after the fire took Greenville, provided $7 million through California's Office of Emergency Services (Cal OES) to support the incoming firefighting personnel from outside the area and police units.

The first real money the county saw was after the settlement between PG&E and five counties in a criminal case brokered by the

Plumas County's district attorney's office. Through District Attorney David Hollister's efforts, $17 million was distributed throughout county organizations. But neither the supervisor representing the area nor the DA had done the grassroots work to determine need and the best organizations to fund. Because of the dysfunction of the county government (an entity with many unfilled and under-marketed positions), these dollars were not distributed in a strategic way to provide the most benefit for the most people. There was an incredible lack of communication and transparency, and this contributed to residents creating their own narrative about who would be receiving funds and who would not. Some organizations sat on astronomical amounts of money and made the process for accessing it difficult. A few organizations, with well-educated staff that could lobby, successfully received better funding than organizations that may have served more people but did not have the ability to lobby successfully. The local school district received money that went into its general fund instead of being earmarked for fire-affected areas.

The funding issues after the fire presented a fact that many were reluctant to see: people in our impoverished working-class community did not have a good understanding of money. Initially, $17 million looked huge. But real recovery would take far more than that.

In our case, it was middle-class families and individuals who least needed direct relief services that were able to return and rebuild. My strongest recommendation going forward, thinking of the many fires to come, is that immediately after a disaster everyone in the community affected needs to pool their resources so they can meet any need that arises. The speed with which people can be served is crucial. Very few have the luxury of waiting days for gas money and funds for crisis shelter and food. I would love to see a national system developed and designed by communities, with training of local people who could then understand and guide the process for immediate and effective relief services. People could do updated training every year or two. That way, when a fire disaster happens—and they will continue to happen—there will be boots on the ground, people who are familiar with the community and its needs, people knowledgeable and trusted who will be able to help people apply for FEMA and other aid. Everyone will have needs after the fire, regardless of income and whether they are property owners or tenants. Fire recovery is stressful and traumatic, but as a community we should be able to make it less so with better planning and more accountability from local elected officials.

Ground Truth
The Limits of Scale
Zeke Lunder

The Calamity Tour

I used to think I knew what was going on in the woods. I've been navigating the worlds of forestry and wildfire since the late 1990s, when geographic information system (GIS) mapping was just beginning to be used operationally on large fires. I've spent my career surveying wildfire hazards with drones, light planes, helicopters, and dirt bikes, analyzing whatever wildfire-related GIS data or imagery I could get my hands on. My coworkers and I have spent the last two decades mapping over 250 of California's largest wildfires, and in 2021 we printed enough fire maps to cover two acres. But when it comes to describing the scale of our wildfire crisis, I feel pretty lost.

In the spring of 2023 Lenya Quinn-Davidson, with University of California's Agriculture and Natural Resources program, asked me to help lead a field tour in Butte County for the national Fire Networks team. The Fire Networks partnership works through three interconnected peer-learning networks—the Fire Learning Network, Fire Adapted Communities Learning Network, and the Indigenous Peoples Burning Network—and supports the Prescribed Fire Training Exchanges (TREX/WTREX). This partnership is supported by a cooperative agreement between the Nature Conservancy and federal land management agencies to support people building better relationships with fire in places across the United States.

The previous summer, Lenya and I had been out to look at the reconstruction of Paradise following the 2018 Camp Fire, and she wondered if I'd be interested in taking a group on a similar tour. I don't really like spending time in the new version of Paradise. Old Paradise had shade and a funky charm. Now it is hot, dusty, windy, and covered with Scotch broom and new modular homes. Being there triggers mourning for both what we've lost and what we're sure to lose again. I'd rather go just about anywhere else. We agreed, though, on the need to subject the group to some of the big carnage around here, not little feel-good local projects.

Over the years, I have noticed that every time a wildfire lies down when it runs into a fuel reduction project, we trot out a congressional delegation to tour the site, write papers about it, or turn the success story into catchy graphics for social media. We rarely tour the places where fuel breaks failed or talk openly about the limits we face in dealing with megafires. This tour would be for people who spend a lot of time thinking about how humans can coexist with fire. After what we've been through in Butte County, I'm convinced there are some places we shouldn't waste our efforts. I thought dropping the group into the middle of an enormous calamity that represents our new wildfire reality might help drive this discussion, and maybe I wanted to share my sense of disorientation and despair too.

Trauma, Language, and Storytelling

We've been through a lot, here in Butte County. In 2017 historic floods threatened to breach the enormous Oroville Dam, and over one hundred thousand people were evacuated. Just up the hill from where I live, in Chico, the 2018 Camp Fire killed eighty-five people and made over thirty thousand people homeless. Thirty of my coworkers lost their homes. For several years, friends and families hosted refugees in trailers and RVs in their driveways all over Chico. Nearly five years later, many have rebuilt, but people are still living in trailers on their properties; others live in tent camps on the edge of town. In the aftermath of the fire, heavy rainstorms pounded the burn, and the toxic stormwater from the fifteen-thousand-acre urban area of Paradise poured into our favorite swimming creeks. To our west, the 2020 August Complex burned a million acres across most of the Mendocino National Forest, choking our pandemic scene with months of smoke. To our east, the 2020 Bear Fire burned four hundred square miles, leveled another fifteen hundred structures in the community of Berry Creek, and killed sixteen people. To our north, the 2021 Dixie Fire incinerated the town of Greenville and blackened a million acres of the lands I know best.

And that's just the past six years. Before that, fires in 2008 threatened Paradise and wiped most of the community of Concow off the map. Many of the people who rebuilt lost their homes again to the Camp Fire. (Concow also had a major fire in 2000.)

When Lenya called, I had just visited Big Bald Rock, inside the Bear Fire scar. It was my first time there since the area had burned, and I was struck with three superlatives: the huge areas of high-severity burn, the incredible density of the pre-fire forest, and the insane rates of regrowth in the oaks and brush.

I've always tried to avoid using loaded terms like "devastated," "catastrophic," or "scarred" to describe burned areas, but recently I find myself using these words, usually prefaced with an F-bomb. Our landscapes and people are scarred and devastated. The outcomes of our recent fires are catastrophic. How else do we describe over two thousand square miles of black-stick forest, endless clearcuts of salvage-logged private land, and neighborhoods devoid of trees, peppered with blue-tarped trailers? It does look like a war zone. I know everything won't be black forever, but our current situation is difficult to describe in neutral terms.

Salvage Logging in the Dixie Fire

We tell stories to try to make meaning out of the situation. Also toward that end, I've created hundreds of maps explaining the cold, simple facts of the Bear, Camp, and Dixie Fires. In a single day, the Bear Fire marched thirty-five miles from red fir forests in the high country to gray pine and live oak foothill scrub at Lake Oroville. The Camp Fire ran eighteen miles in twelve hours, destroying almost twenty thousand structures. The Dixie Fire's biggest run consumed over one hundred thousand acres in two days. I enjoy teaching and talking about land, but lately I feel maps fall short in truly communicating the staggering scale of change in our forests.

One map shows the Camp Fire, about fifteen miles wide, about four hours after ignition. In a strictly technical sense, the map is a great teaching tool about the limitations of fuel breaks during extreme weather, showing the fire spreading miles ahead of itself via long-range spotting. On a more human level, it utterly fails to capture the true horror of the scene. Yellow dots in the center show burning buildings. All of the major roads out of town are blocked by fires, and thousands of people are trapped inside this scene. Dozens have already perished. A regional maternity ward is on fire. Women are going into labor inside burning ambulances, firefighters are trying to save the hospital, parents at work in the valley can't make it up the hill to pick up their kids from school or save their pets.

So many stories in one map. Looking over Chester, California, a map shows over seventy miles of contiguous fire, a burned town, and a part of the Dixie Fire that, in two days, burned over 35 percent of the more than one hundred thousand–acre land base of a timber company that has been heralded as a leader in sustainable forestry. At one point in the summer of 2021, almost every citizen in Plumas County was evacuated or under an advisory evacuation.

Pace and Scale

The California Wildfire and Forest Resilience Task Force has a stated goal of "treating" a million acres of wildfire fuels a year by 2025.[5] But what is our end game? Are we hoping to make every at-risk community safe from fire? Do we think we can manage enough forest land to keep the next Dixie, Creek, Rim, Caldor, Camp, Carr, Ferguson, or Bear Fire from happening? Right now, there is a lot of interest in expanding the scope of prescribed fire, but the vast majority of acres being funded are thinning and mastication projects. ("Mastication" in this case refers to reducing forest vegetation by grinding or shredding.) There is no business case that suggests we can process and dispose of a million acres of masticated woody biomass annually (the entire California timber industry has logged about one hundred thousand acres of private land per year over the past twenty years), and momentum toward building new utility-scale biomass cogeneration plants is slow.

Large-scale fuel treatments can have forest health and public safety benefits, and they are critically important to accomplish adjacent to our forested communities, but California's geography limits the amount of acreage we can mechanically thin. On steeper, inaccessible slopes, prescribed and managed fire are our only real tool for reducing fuels at the landscape scale. Despite a lot of talk about it, the US Forest Service is showing very little interest in using fire at the scale needed to really make a difference, and California's industrial timberland owners barely burn anything. Most of them don't even burn piles anymore. If we have more extended droughts, forest thinning projects alone aren't going to stop megafires. We can't mechanically thin the steep ground where the likes of the Mosquito, Dixie, Carr, and Creek Fires grew large.

The assumption that large-scale fuel treatments can affect the outcome of megafires is rooted in all the success stories we have spread after forest thinning projects helped us control a wildfire. But timberlands comprise only a fraction of the wildlands where money is being spent on fuel breaks. In places with fast-growing brush, there's very little lasting benefit to investments in large-scale

---

5. For more information, see California Wildfire and Forest Resilience Task Force, https://wildfiretaskforce.org.

backcountry fuels reduction. Take the example of the 2002 LNU (Sonoma-Lake-Napa Unit) Complex Fires.

Between 2015 and 2019, wildfires reduced fuel loading across about one-third of the 1.5 million–acre area. Yet when the six-hundred-square-mile 2020 LNU Complex burned, destroying close to fifteen hundred structures and killing six people, it ran across over two hundred square miles of land that had burned within the previous eight years. One hundred square miles had burned just two years before. Just about anywhere we cut brush will need treatment again very soon, and forever is a long time. Are we going to keep asking taxpayers who live free from wildfire risks in Stockton or Palm Desert to pay for this work in perpetuity?

The Planning Trap

I've found a career niche working on "landscape-scale wildfire hazard assessment" projects. We use satellite imagery, fire history maps, parcel boundaries, roads, terrain, weather data, and other map layers, along with our experience working on large fires to identify priority locations for large-scale fuel-reduction projects.

So many things about planning work feel good! Modeling fire behavior and building GIS databases is not much fun, but drawing shapes on a map is cathartic—it feels impactful—and the solutions seem very clear. We'll put in a ridgetop fuel break here, do a prescribed burn there, throw in some mastication and pile burning around the "assets," and everything will be good to go. And there is the fieldwork component: we get to learn new landscapes, drive or ride bikes around in the woods, hike, look at past fires or logging projects, and share stories about land with other passionate people. For the most part, though, very few of the large-scale projects I've designed in my career have actually been implemented.

In 2019 and 2020 the company I work for, Deer Creek Resources, had funding from the federal Bureau of Reclamation and the Sierra Institute to develop a conceptual plan for where to best thin and burn across the 850,000-acre South Lassen Watersheds Region. Since there was a pandemic going on, I was able to take my eight- and ten-year-old sons with me for the survey portions of the project. We escaped the choking wildfire smoke in the valley and hiked into roadless old growth stands near my hometown of Westwood, shot big sugar pinecones out of tall trees with a .22, hung out in hammocks, and camped near creeks. We got to explore places I've always wanted to see.

But within months of finishing our plan, about half a million acres of our project area got roasted by the Dixie Fire. Our models of fire effects greatly underestimated the observed fire behavior. Many of the coolest old-growth areas we had visited were destroyed, and areas we'd thought were in pretty good shape (thinned and actively managed) also got nuked. The Dixie Fire was devastating in so many ways, but the hardest part to swallow was the realization that even when we think we have technical solutions that might save our forests, the scope of our forest health problem is far beyond our ability to manage it. What made us think we could?

We talk a lot about the Forest Service's dwindling capacity, but the problem is more fundamental. Fire is the only tool that is really up to the job of managing fuels at the landscape scale, and the Forest Service is first and foremost a fire suppression outfit. As far as their land management capacity goes, they ramped up for about forty-five years after World War II, to carve about four hundred thousand miles of roads into the backcountry and drag the largest trees off our western landscapes. They followed this up with a flurry of tree planting but not much tending. Then they got in trouble for all the bad things they had done, and we sent them to planning jail, effectively shutting down any activity for thirty years.

For my entire career, the Lassen and Plumas National Forests have been involved in one landscape-scale planning effort after another. Millions of labor hours have been spent skillfully designing incredibly detailed projects to increase wildfire resilience, yet in the past fifteen years the landscape has lost more than half of its trees. The Dixie Fire burned through proposed uncompleted projects that were funded out of settlements from fires that occurred as far back as 2000 and 2007.

The West is littered with relics of booms and busts. Forest Service offices are the latest ghost towns. A huge number of the trees they planted have turned to ash and smoke, and their amazing forest road network has been blown to pieces by storms and neglect. We have equated the ability to remove all the big trees with the capacity to manage ecosystems, but nobody really even knows how to do this. Wildfires are eating foresters' lunch, regardless of their employer.

Learning the Limits

Maps and data visualization can help us make sense of numbers and trends, but even the best require field-validation. Without proper context, our ability to conceptualize problems and plan appropriate solutions can become completely disconnected from reality. This becomes crushingly clear when you stand in Berry Creek or Concow, realizing that everything you can see in all directions has been burned, and that there's clearly no technical solution, either to preventing the next fire or "fixing" the damage from this one. That's what the field trip was really about—ground truth.

People are small, forests are vast, mountains are rugged, logging is brutal, and economics are real. We can absolutely accomplish projects that will make our communities safer and forests more wildfire-resilient, especially if we scale up our use of fire, but the megafire train left the station as soon as we started cutting down all the big trees and suppressing every fire. We are one hundred years into this passage, and we'll lose millions more acres of forest before the reset is complete.

Big burns are traumatic, and the darkness is real.

Triptych in Smoke
Lasara Firefox Allen

Smoke One: A Bloody Sky (**Lightning Complex Fire**)

There is no sun. No light. Just a blood-red, technicolor sky. We wake up to what feels like it might be Armageddon. We find there is no internet. No cell signal. We're isolated and cut off from any word of what might be at hand.

My Gen-X, nuclear-shadow-panicked mind goes directly to nuclear annihilation; do not pass go, do not collect two hundred dollars.

Is it safe to go outside? Will my skin melt? Will my cells mutiny?

I decide that if this is nuclear winter, we're already as good as dead. I go outside.

I do not melt. But the air smells distinctly of a heavy, campfire-like smoke. And there's ash coming down like a fine snow.

Still no cell signal. I quickly gather the family, get the cat into his carrier, grab the bug-out bag—which holds a small, portable, multi-station radio, our most important papers, and a few pieces of clothing for each of us—from its spot by the door. I load up our laptops and a couple of boxes from the closet holding mostly photos, then add some basic survival supplies from around the house (blankets, a knife I always take when I go camping, some food) to the pile in the cargo space of the car. After we remember one more thing, a final quick hunt produces everyone's required medications, which are thrown into a bag and tossed in the back of the loaded car. We climb in and head to town.

Residents of a Northern California wilderness enclave, we live in a surprisingly overpopulated tinderbox with a single road for ingress and egress. Still, thankfully, it's early, and there aren't many cars on the road yet. We get to the nearby town, Willits, a seventeen-minute-drive, in thirteen minutes flat.

In town, it's worse. The ash is falling thick and gray. I lift my phone, and there's still no signal.

I get out of the car at the gas station to ask folks if they know what happened. The ash gets in my hair, my eyes, and my mouth. It's gritty and tastes of vaporized plastic, destroyed grasslands, and a million burned wildflowers.

The trimmigrants (drifter kids who work on the local pot farms) have panic in their eyes. One holds up a hastily scrawled sign that says simply "HELP!"

Near an AT&T van, I see a couple of guys in reflective vests.

"Hey, do you know what's going on?" I ask.

"Everything between here and Santa Rosa is burned," says one guy, meeting my eyes as he ashes a cigarette on the concrete at his feet.

Still no signal. No way to verify this word-of-mouth report.

We wait in line at the gas station to fill up the car. As we do so, a pair of terrified-looking young adults and their dog approach our car. "Hey, can you get us out of town?" asks the older-appearing of the two, voice wavering a little.

"We're full," I say, gesturing to the cargo space piled high with some of our most precious possessions, and the kid and the cat taking up the back seat. "Sorry."

By way of apology, I hand them some food and a few bucks. They take it and turn away, shoulders tense and rigid under their dirty denim vests.

Car gassed up, we head embark on the forty-five minute drive to Mendocino and the coast. We have no idea what we'll eventually be coming back to, but at least by the water it's more likely we'll be able to breathe.

Smoke Two: A Cathedral of Tree and Air (The Camp Fire)

My friend Randy and I have planned a hike day for weeks, but when I wake the air is soupy and thick. I text: "Hey Randy! Are we still gonna walk?" We decide yes, so I put on my ever-at-hand N-95 respirator mask, grab my keys, hop in my Honda CR-V, and drive the couple of miles to the trailhead.

After a hello hug and greeting, Randy and I take to the trail. We hike down toward the waterway and into the trees. At first sparse and spread out, the fir stands will give way shortly to more dense groves of redwood. As opposed to sunny-day banter, we walk in relative silence, the smoky pall too heavy to lift.

We approach the arched gateway made of trunks and branches and enter into the grove. The air, by the miracle of redwood magic, is clear, cool, soothing. I whisper a reverential prayer to the ancient giants and remove my mask for a moment. The deep, rich scent of redwood humus and the airy fragrance of bark and needles mix in my rejoicing airways.

It's all I can do not to cry.

Smoke Three: Smoke Season (Deep Fire, South Fork Complex, Lightning Complex, Happy Camp Complex, Smith River Complex)

When I was a kid, the season was called Indian Summer. It's that time in the fall when the days are starting to grow shorter but the season, at least in California, is still hot as hell.

I'm living in Oakland now, years and miles away from the arid hills outside of Willits. And, sure as death or taxes, the season rolls around, only its new name is Smoke Season. I open my air quality index app and see how safe the air is to breathe, whether it's all right to go outside or whether we should keep all the doors and windows shut. My mouth tastes bitter, acrid. Stale smoke is even worse than fresh.

The AQI is red. At least it's not purple.

I grab my mask and get ready to head to work. This is the new normal.

The doors and windows remain closed. They will remain that way even tonight, when the cool night breeze would be the only relief from the stultifying heat. But until the AQI improves to yellow at least, the windows will stay shut tight.

I get to my office. It's in an old building with rickety doors and windows. I'll be wearing my mask all day, due to the dual specter of COVID and smoke seepage through the misfit frames.

Hours later, after a full day of talking and breathing through my mask, my face is raw and red from rubbing against the fabric, and my nose and chin are speckled with "maskne"—a recent-day minor plague. I'm dehydrated and hangry because, as the boss at an organization in post-pandemic crisis, I can rarely find the time and privacy to remove my mask and take a drink of water or a bite of food. My ear cartilage is sore from holding the elastic bands all day.

At least we have N-95s, I think, remembering those months in early COVID when people were sewing piles and piles of cloth masks. Maybe those cotton remnants saved some lives then, but they'd do little to protect lungs from this carcinogenic soup.

It's the end of my work day, and I head out to my car in the oppressive afternoon heat. I start navigating back home, lights on for visibility although it's barely past 4:00 p.m. At a traffic light I see an unhoused neighbor sitting by their tent, masked up, and whisper a quiet prayer of gratitude for Mask Oakland, a local community aid organization that started distributing masks to unhoused communities in the East Bay in 2017 and has continued doing so during fire season and in major smoke events since. They were ready for COVID.

The person and I make eye contact, and I offer a brief nod and a half-smile. I can't tell if there's a smile in return since we only have half-faces these days. Denser smoke drifts in, making my neighbor seem less substantial. I drive on.

Bringing Fire Back to the Land

An Interview with Margo Robbins of the Cultural Fire Management Council

Dani Burlison

Margo Robbins is a Yurok Tribe member who lives on the Yurok Reservation along the Klamath River. She is the cofounder and executive director of the Cultural Fire Management Council, a community-based nonprofit organization that hopes to bring traditional fire practices back to the Yurok homeland so that it can once again support the traditional life ways of their people.

Their service area and focus is the ancestral territory of the Yurok people, which stretches across California's North Coast, including much of coastal Del Norte and Humboldt Counties. The council conducts burns on tribal and private property for both tribal and non-tribal people in the region. Margo is one of the key planners and organizers of the Cultural Burn Training Exchange (TREX) that takes place on the Yurok Reservation twice a year. She is also a co-leader and advisor for the Indigenous People's Burn Network (IPBN), a support network for Native people who want to bring back their fire cultures. The IPBN targets tribes across the United States and Indigenous nations from other countries, and the work is facilitated by the Nature Conservancy. Margo was a part of the original planning for IPBN and collaborated on creating a healthy country plan called the Yurok-Hoopa-Karuk Healthy Country Plan, which outlines a way to get from where Indigenous people are at today with fire, back to true traditional burn practices.

DB: For people who maybe haven't heard the terms "cultural resource management" or "cultural burn," can you explain what those mean?

Margo Robbins: Cultural burns have to do with who is doing the burning, why, and when. Cultural burns are led by the Native people from that place; it is very much place-based. The people have an intimate connection to the land and knowledge and insight about the elements that affect fire. Cultural burns have to do with using fire in ceremony and as a land management tool. Traditionally we depended on the land to provide everything we needed; food, medicine, clothing, homes—all of that. And the utensils and tools to go with that. So cultural burning has to do with increasing the health and availability of those culturally important plants and animals, as well as restoring the land to a healthy balance, whether or not the individual components of that are personally useful to us. So it's not all about us.

DB: What other reasons, besides strictly fire mitigation, would someone engage in and help teach the practice of a cultural burn?

Margo Robbins: Cultural burns are not about fire mitigation. They are about cultural resources and healthy lands. A by-product of that is wildfire prevention. So just like a by-product of prescribed burns is cultural resources, people that live in that place do and can go out and gather.

DB: I feel like most people around the state think that if there's any kind of planned or managed burn, it's simply done by the fire department, or it's just for preventing wildfires. So I'm glad that you made the distinction between that and the cultural burn. Can you describe what the actual process is? I know you mentioned a little bit about ceremony. What is the actual process of engaging in a cultural burn?

Margo Robbins: We do a couple different types of cultural burns, and one does not look like a cultural burn at all, because we conduct it as a training exchange for qualified firefighters. And all the people there are dressed with hard hats and yellow shirts and green pants and fire boots. We have fire engines and hoses. So what makes it a cultural burn . . . is that the places that we choose to burn (are rich with cultural resources) and the timing of the burn. So we are burning plants in particular stages of their growth. Like, for hazel, we want more hazel in it—we want straight new shoots. So there are particular times of the year that we burn it. That is one of the ways that we're able to burn larger pieces of land, like maybe thirty to forty acres in a day. And so we do those cultural burn training exchanges twice a year, once in the spring and once in the fall.

DB: And is that the TREX program?

Margo Robbins: That's the TREX program, yes. For that, we have a crew of workers that cut a fire line around the areas that we intend to burn. The fire line consists of cutting all the brush in a ten-foot-wide swath and then, within that ten-foot-wide swath, scraping it down to the dirt two feet wide. That would go all the way around the burn unit to keep that fire in the place that you intend for it to stay. And then we have about thirty-five people come for the burn at all levels of expertise and experience. Some have decades of experience—they're incident commanders and burn bosses and firing bosses, all of the different qualified positions. And some of the people have never been on a fire before in their life. They learn how to burn. They have taken either an in-person class or online, what they call Basic 32 classes, and it's entry-level qualifications. And then they have to do a pack test, which is to pack forty-five pounds for three miles in forty-five minutes. And they have to do some fieldwork, like digging lines and things like that, which they do here at TREX.

So once everybody is here, we will orient them to our landscape. We'll go to a cultural site to give them an understanding of why we burn—that it's not just burning brush, that it has to do with the continuation of our culture. We will go out and walk the burn units that we plan to burn, and we will point out the culturally significant plants to make sure that they know how to identify those. And we will also talk about potential hazards, potential firing patterns, and talk about the burning of the unit while also teaching them about the cultural importance.

And on burn day, we have a group of fire lighters and a group of fire holders. They start firing on the top of the unit. We will typically have the test fire at the highest point, and we will send up prayer and use wormwood torches to start that initial fire.

And then we carry the fire down the hill in parallel lines to the hillside, going back and forth, laying fire on the ground. So that's one of the ways that we do cultural burns. The other way is a more traditional style of burning, and that's family burns.

Traditionally, families took care of their own hunting and gathering places, which started near the edge of the village because they made their villages in places that had a lot of food sources. Basically, they're starting to burn near the edges of their villages. Again, the by-product is wildfire protection. So, anyway, for family burns, we actually do a one-day class that we encourage people to come to because our lands are so overgrown with vegetation now that we want to be cautious about burning, and we want other people to be cautious about it too. We can't just go out and burn in the right place at the right time anymore because it's so overladen with fuel, and the wildfire risk [is high]. So, because of COVID, we haven't been able to do our class. But it's basically about either reminding or teaching some people for the first time about how to burn safely.

And then we have tool lending. People can borrow tools to use to get their land ready to burn. We have a couple of people that can go help them. Because it's a lot of work to either cut a line around what you want to burn or cut the brush and pile it, depending on what people want to do. And then the day that they want to burn, we bring what we call a slip-on fire engine. And really it's a mobile source of water that can be pumped at high pressure in the back of a truck. We bring that for a safety precaution.

People of all ages can burn during family burns. They do it during the time Cal Fire doesn't require a permit, which is sometime in October through the end of April. We have all ages of people, from babies being carried on their mama's backs or fronts, to two-year-olds learning how to light under the close supervision of an adult, to elderly grandparents out there.

And they are burning smaller acreage, usually an acre or less. There's a few that will burn more for their hazel-gathering places. One of our elders has about five acres she burns. And so that is the family burn program, which is a cultural burn.

DB: I'm curious about the cultural burn training exchange, the TREX program. If you are working with firefighters, is it usually local fire departments or Cal Fire or a combination?

Margo Robbins: For TREX, we have people that come from all across the United States and other countries. And people come from Spain and Ecuador and Canada, and it's also local fire people. We have about a dozen Cultural Fire Management Council people, CFMC people, that help with the burn. The Yurok Tribe has about maybe five people that help. Sometimes our neighboring tribes, the Hoopa and Karuk will send down a few people. The Hoopa Tribe actually had at least one of the burn bosses qualified in our TREX. And then it's a combination of agency and organization and private people. We'll get people from BIA, BLM, Cal Fire, Forest Service, different land trusts.

We usually . . . have at least one or two individuals that are just interested in learning how to burn because they perhaps want to take care of their own land with fire. And the important thing to remember about TREX is that it is the participants that are teaching each other. And that's why they call it a training exchange.

DB: I understand that for many, many, many, many generations, Indigenous people in California and other places have been doing these cultural burns, and there was a time period where this was stopped. I'm curious if you can talk about why that was. And again, this is one of those obvious answers, but just from your perspective, I'd like to hear from you.

Margo Robbins: I think it was because the non-Natives that came to our land were afraid of fire, and they didn't understand the benefits of it.

DB: Right, so their perspective was mostly about just putting out the fires after they had started instead of this relationship with the land and using fire for it.

Margo Robbins: Yeah. There were the ranchers who used to burn. They saw the Natives burning, and they did realize that it produces healthy food for the grazing animals. And so the ranchers used to burn, and I don't know if they still do or not. I think that's maybe something that's coming back for them as well, because they were out on the land and trying to live in a . . . maybe you could say less abusive relationship with the land.

DB: They weren't just out clearcutting and bulldozing everything.

Margo Robbins: Right. They actually paid attention to what the Natives were doing to keep the grazing lands healthy and open.

DB: I just have this image in my head of the settlers, the ranchers peeking over the fence to see what the Indigenous folks are doing. And borrowing some of those practices to benefit themselves.

Margo Robbins: Exactly. Exactly.

DB: How and when did this practice start up again? I'm curious about what the process was like and how it was maybe received by different agencies or government or authority in these areas.

Margo Robbins: It never completely disappeared. The knowledge was passed out through the generations, and there were those who were willing to risk the punishment for starting fires, and they did continue to bring to some degree . . . illicit burning. Of course, it wasn't nearly as prevalent as it had been in the past, but it did remain to a small degree.

I'm not really sure in other places how that went. I can only speak for us, here on the reservation, that brought back our fire culture and how that happened, but I certainly do not speak for the tribe as a whole or any other tribes or any other place. How we did it is we recognized the need for basket weavers to have hazel. We must have hazel in order to have baskets, and baskets are at the core of our culture.

And also we were worried about wildfire. We were part of the California Endowment Building Healthy Communities initiative. And we got a small grant from them that connected fire to health, the health of the land and the health of the people, and to move up some sort of plan in how to do it, but we didn't wait to get a plan.

We knew that we needed fire, and we needed fire now. So we did some research to find out, "Okay, who holds the power? Who has authority over fire, and what do you have to do to be able to burn?" And, as a community, we sought the support of the formal tribe. And once we found out who holds that power, we had a public meeting with media there. Getting them to publicly say, "Yes, they support the fire."

But of course they wanted to bring fire back. It's actually written into our constitution that we'll use fire. And so that was not a hard sell. And so our very first burn that we had, we got Cal Fire to come do it, and the Yurok Tribe, and they burned seven acres.

It was a traditional hazel-gathering place. The community cut the fire line around it. We didn't have any money. I just made deer meat sandwiches lunch for them and got out there and worked alongside them. And that goes a long way. We cut the fire line, and then Cal Fire came in. It was an inmate crew that they brought. And then I think there was maybe four or five Yurok tribal people. And when they came, Cal Fire with their inmate crew, I counted how many of them came, because I wanted to know, how many people do we need to burn our own lands?

And there were twenty. And so I set my goal to train twenty community members with formal firefighter qualifications so that we could burn without going to jail.

And so that's what we did. The next year we were introduced to the Nature Conservancy, and Jeremy Bailey [fire training and network coordinator at the Nature Conservancy], and he taught us how to do TREX. He came, and he pretty much did all of it that first time, but he trained us over a few years and [taught us] how to set up our own TREX. And the Nature Conservancy has just really been our solid guidance support.

We started out our first TREX with a training open to the community so they could get their entry-level firefighter qualifications. And the whole community center was full of local people. It was so awesome, my son among them.

We went quite a few years without money. The people that were trained for that first TREX, they have to support their families, and so some of them went here and some of them went there. Some stayed around, but they weren't able to just commit to burning with us because it takes time. They need to be able to support their families. But I don't even know how many years ago it's been now that we hit our twenty-person goal of qualified firefighters. I think our first TREX was 2013.

I would like to add just a little bit about smoke. Smoke is a factor that can be prohibited in some places. It's not here where I live, because everybody accepts and wants to have fire. But nobody wants to live in a tinderbox. And they see the beautiful results of the fires that have taken place, and so they want and accept fire and the smoke from fire. But in some places, people don't want to have that smoke, and they call and complain that the air quality district has issued smoke permits when they've just gotten done breathing wildfire smoke for a week or two or three or four, but the truth of the matter is that prescribed and cultural burns reduce the spread and intensity and likelihood of a wildfire.

People can either choose to accept the minimal amounts of smoke from cultural and prescribed fire, or they can continue to breathe in this massively toxic smoke for weeks on end from wildfire. But they have to make a choice, because there is no such thing as a no-smoke option.

DB: I'm so glad that you brought that up because I was thinking about that before, when you mentioned the specific times of year when you don't need a permit because it's a smaller chance of wildfires starting. And it's true, I'd much rather have those burns than the smoke that we get here in Sonoma County when there are hundreds of houses burning down. That's so toxic.

Margo Robbins: Right?

DB: It's so toxic. And with these burns, you're just burning the underbrush, right? You're not losing entire trees and homes and buildings and cars and all of that.

Margo Robbins: Exactly. Exactly. And these wildfires are feeding on the massive amounts of vegetation and dead, woody debris on the forest floor. So, if we remove that with cultural and prescribed burns, then it's not going to have as much to feed on. And it's not going to grow into these megafires like it currently is.

The other thing that I would like to say is that, as I said earlier, at one time it was families burning their own hunting and gathering areas. And fire belongs in the hands of the people. The average person should have opportunities to learn how to burn safely, and they should be able to use fire to take care of their land and protect themselves from wildfire. You just think of all the private landowners. If you have ten thousand people that own one acre, that's ten thousand acres burned in a year. If you have a hundred thousand people that own five acres, we're going a long way towards wildfire prevention. When we look at the numbers, if we will just allow people to have that basic right of using fire again.

DB: I could see it here. Looking at all the places that have burned here in Sonoma County in the last several years, if there were more of the prescribed burns happening in certain areas, I could see entire neighborhoods would still be there.

Margo Robbins: Exactly.

DB: I'm curious about how you integrate this traditional practice in these modern times when there's usually so many steps and permitting processes and things like that. And a secondary question is: how has it been for you personally to be really ingrained in this process, this revitalization of this practice that's so culturally important to you?

Margo Robbins: It has just been such an amazing journey. It was so gratifying this spring when the places that we burned last spring came up thick with hazel. I was up on the mountain picking because I'm a basket weaver, and here would come car after car and even a vanload of kids. And they would come up on the mountain to gather their hazel. And to hear the kids saying "Oh, yeah, I want to pick big ones because I want to make an eel basket," or "I want to make a baby basket," or "What size should I pick for a rattle?" It's such a wonderful feeling to just hear those kids up there all excited about the hazel and to just see so many people coming to gather the fruits of our labors, like, "Yes, this is why we do this."

It is such a wonderful thing. I remember that first little unit that we burned, and there was just the seven acres. And we felt like we really had to be misers with our sticks, because there were just not that many in the place. And it was like a burst entry burn—very quick—so it didn't burn really clear. There was still a lot of brush in there, so we couldn't get to a lot of the sticks that were there. And so, if I saw a car stopped there, I would get out and make sure that it was a basket weaver that was picking because we did not have enough for people who were going to go in and pick and try to sell to basket weavers.

DB: I didn't even think about that.

Margo Robbins: Yeah—well, we never did either. And then suddenly it's like, "Oh my gosh, we've got to guard our sticks." But now we are able to burn so many acres that we don't have to worry about that. We can invite our neighboring tribes to come pick with us. We can be happy when we see the vanloads of people pulling up to pick hazel. And if the occasional person is there to pick and sell to other people, then I realize now, well, not everybody can get up to pick sticks, and so they are providing a service. What they're doing with the money, I don't know, and I'm not questioning them.

But they're all being reconnected to the land, and that's so critically important. And if we can just reconnect our people with the land, it is going to be a huge step towards leading us back into a healthy way of living. And for those who are able to come out and participate in the burns or even just observe, it has a way of drawing you in and bonding you to the living things around us.

DB: What are the potential long-term impacts, both culturally and ecologically, of bringing cultural burns back?

Margo Robbins: Oh my gosh, what we've started is going to carry on into generations to come. There is no way that we will finish the job in our lifetime, and so we are busy educating the next generations as well. Because, really, this effort of bringing fire back to the land and reconnecting people to the land has the potential to restore not just the land back to health but the people back to healthy ways of living and thinking as well. And so the enormity of it—I don't even know the words of it. It's big.

The Little Sawmill That Could
Margaret Elysia Garcia

The rising price of building materials was already a significant obstacle to building in the Sierras—or anywhere in California, for that matter. Various contractors of existing projects in Plumas County were already reporting increases in lumber prices long before the Dixie Fire of 2021. Those whose properties burned report that replacement materials sometimes were double and triple what they were before the fire—even a few months before for places that were under construction. Insurance claim payouts, of course, did not reflect this new reality.

In Taylorsville, California, the Sierra Institute for Community and Environment, a nonprofit organization with a mission to "promote healthy and sustainable forests and watersheds by investing in the well-being of diverse rural communities and strengthening their participation in natural resource decision-making and programs," had purchased a parcel in nearby Crescent Mills along the railroad tracks nearly a decade earlier. The parcel was a former mill site turned superfund site that had sat empty except for heavy metal contaminates for many years. In 2019 the Sierra Institute did a major cleanup of three acres of the site with the intent to use it as a cross-laminated timber plant. But that idea got a new mission post–Dixie Fire. The wood products campus is now the site of the first new sawmill to open in California in three decades.

Nearly a million acres burned in the Dixie Fire, and that translated to roughly 60 percent of Plumas County. The Park Fire of 2024 and other fires in the few years in between have brought that number closer to 75 percent. For residents of Greenville, Indian Falls, and Canyon Dam, the thought of rebuilding, with or without payouts from insurance companies, remains daunting with the risen cost of lumber, which hit record highs during the first year after COVID.

At the same time, there is a burnt-out forest to contend with that has many hazardous trees needing to be felled and dealt with. The two largest mills in the northeastern California—Collins Pines Company in Chester and Sierra Pacific Industries in Quincy—indicated that so much of their own forest holdings had burned that the companies projected it would take them the next five years to process their own timber from just the Dixie Fire. There have been fires in Plumas, Tehama, and Butte Counties since then that have only added to their own private holdings that will be logged. They wouldn't be able to pick up the milling of lumber on private property and national forest lands for years. The Sierra Institute recognized the need to do something about all the dead and blackened trees and heard the call from residents about their need for building materials.

At the same time, watching green trees being chipped near Indian Falls because they might be a road hazard someday felt like insult added to injury for many residents who sat in long traffic lines on Highway 89 approaching scenic byway State Route 70 on the road to Quincy. Supervisor Kevin Goss had many constituents express concern for the fate of trees, especially knowing of the need for building materials.

"With tens of thousands of trees burned, what better way to use them than to rebuild the town of Greenville," said Sierra Institute executive director Jonathan Kusel.

The Sierra Institute purchased a sawmill that, in their words, would "change the dynamics of Dixie Fire restoration and forest management across Plumas County." The Sierra Institute is partnering with longtime local partner J&C Enterprises, which now runs the mill and provides jobs for local workers.

"The lack of lumber markets is one of the biggest barriers to getting any sort of forest treatment done on the landscape," said Camille Swezy, operations forester with J&C Enterprises, a fourth-generation, family-owned logging company based in Crescent Mills. J&C Enterprises was also involved in post-fire cleanup of hazardous tree removal on burned properties in Indian Valley and in downtown Greenville where the fire struck hardest.

The mill so far has brought jobs to the community and ensured local lumber availability. Importantly, it will continue to provide lower-cost lumber to the community, reducing the burden on residents who might otherwise not be able to afford to rebuild. The lumber is not graded to be used in houses for framing, but it is for fences, outbuildings, and some interior structures like cabinets and railings.

While J&C Enterprises has historically worked on a wide range of projects, from fuels reduction to timber harvesting, they have in recent years struggled to move timber to markets given the lack of availability of sawmilling operations and purchasers.

Since the 1990s, timber-processing capability has been in decline across the state, with increasing concentration of industrial-scale infrastructure. However, in recent years, the decision makers and policy makers in the state have been looking at "small-scale businesses as a necessary part of the infrastructure required to restore the state's forests," said Kusel before the opening of the mill. "Sawmilling capacity is a key piece of the puzzle that has been missing, though for Indian Valley this will be vital."

To do this project, the Sierra Institute was able to redirect grant funds coming from the Sierra Nevada Conservancy to purchase the mill equipment. The Sierra Institute had been hoping to recruit businesses in the coming years to manufacture wood products at the site. These efforts quickly moved forward as a result of the Dixie Fire.

Since the opening of the mill in 2022, property owners have been able to have their trees processed at the plant and turned to boards. Had they simply let FEMA and Cal OES clean up, the dead trees would have just been removed without anything coming back in return, according to the institute as well as many in the community.

The mill also leaves end cuts and scraps at its entrance on Highway 89 for nearby residents who need firewood in winter and cannot afford it. The mill announces on social media from time to time what they have for residents: eight-foot and sixteen-foot lumber, cedar this week, pine the next.

At the ribbon cutting in 2022, people from near and far came to wish the project well. The site was already a welcomed and celebrated venture in the rebuilding of Greenville.

The Sierra Institute continues to explore other potential partnerships that might model how rural, forested California can build and rebuild homes and strengthen communities facing the continued threat of catastrophic wildfire.

"We recognize that innovative wood products, especially mass timber, can be fire-resilient and represent a real possibility to remake a community with hardened homes to address tomorrow's risks," said Kusel. All around town, the mill's cross-laminate timber products can be seen in the new construction of at least half the residences coming back to Indian Valley.

That J&C Enterprises is the partner in the venture is no surprise to residents of Indian Valley. Jared and Cindy Pew are the younger generation of a four-generation timber harvesting family. Their ideas about logging have evolved since their great-grandfather first felled old-growth forests in Indian Valley at the beginning of the twentieth century. In the late 1990s the Pew family business was one of the first timber harvesters in the area to focus on forest health and stewardship, with thinning projects aimed to help reduce the risk of large wildfires. Jared's father, Randy Pew, a fixture in the valley, won awards like West Coast Logger of the Year, Regional Logger of the Year, and Sierra Cascade Logger of the Year for his conservation-minded practices. Jared's brother Tyler chaired the [Re]Build Greenville Committee of the Dixie Fire Collaborative. An architectural designer by trade, he focused on recycled materials and using off-cut material that others might discard. His designs have long incorporated and used recycled and off-cut materials in designs of parks and buildings. He often champions design students at his alma mater, the California College of the Arts, who have had a continual presence in Indian Valley since the fire and have helped come up with ideas and plans that residents can use to reimagine and rebuild downtown.

The Pew brothers have very different personalities and viewpoints. When you talk to them, it's hard to think that Tyler—a San Franciscan till 2021 when he returned to help the rebuild after the Dixie Fire—and Jared, the brother who stayed close to home—have much in common other than the love they have for their parents and the valley that raised them. Yet both are working now not just for the survival of the community but also for the growth of the valley, each in his own way. Their causes, like those of the Sierra Institute, are now intertwined, forged in fire.

We Circle Up for Fire
A Community Burn
Redbird Willie

Footsteps crunching on the ground bring forth the smell of dry pine duff. Cedar boughs swaying warmly in the sun float flowery scents in the light breeze. There is a quiet excitement in the air. The morning is looking just right. We are optimistic. In the last couple of days there has been a steady wind, but it came up at the end of the day, a late-afternoon easterly. The temperature has been cooperating, climbing at most to the upper seventies. Geese braying alternately in loose formation, pass over, skirting the treetops. They ask:

"Is this going to be the day?"

"Is the land ready?"

"Are the people ready?"

"Yes, I think this is the day."

Months of preparation have gone into this day. At first there was the question: should we even do it at all, with all those people? At this time of year (early May)? The last couple of times we did it, there were less people. Three years in a row we had done it. With each burn the numbers had gone up slightly, but there were never more than thirty. But now two hundred people? Three hundred people? Are we crazy? That question was quickly answered. Yes, we were crazy. But yes, we can do it—it will be great! More participants would be even better. It will be a true community fire. Just like the old days.

First we needed to pick our community burn boss, that special person to bring it all together. We knew who it could be. We asked her. She could not refuse. The job was designed for her. It was a special offer, the perfect job for her in our organization. After a few days to think about it, even though she already knew the answer, she told us yes. She would bring together the leaders, the knowledge holders, the distinguished guests, while also dealing with permits, logistics, communications, schedule, and infrastructure.

Our ancestral skills gathering is always in the first week in May: Beltane, cross-quarter day, the middle of spring in our watchful seasonal eyes, in the eyes of the land. And although there had not been much rain this past year, there had been a few winter storms. There was just enough to bring spring to the land. California plants can do with very little water. Many can go two hundred or more days without water. Most times, though, all they need is a reminder of water: "Oh yeah, water! Thank you!" Despite the low amounts of rain, it was green everywhere. Cream-colored iris and purple and white lupine blooms were sprinkled throughout the camp, super bloom years for both. It was a California spring.

So the preparations began. There were many questions to answer. Do we get a permit? Shall we get an official by-the-rulebook burn boss? Or would an Indigenous elder be good enough? This is the quandary we always live with as Native people, struggling to decolonize and revive our Indigenous ways. Do we follow the new way, the mainstream, new rules, patriarchal, top-down path? Or do we follow the path that has worked for centuries, the elders, spirit, women, and men in a jointly led community path?

In these times we end up doing a little bit of both. We follow the mainstream rules to stay out of trouble, but we also have this guiding hand of Indigenous, ancestral leadership just under the surface. Well, sometimes just under the surface and sometimes right out in front for all to see. It depends on the circumstances. The old path we continue to cultivate and nurture. Someday it will again be the main path. This new path that struggles with its heart is unsustainable and will fall away. In these times we keep the good path alive with our thoughts, our memories, and our stories.

Back at the gathering, we are also planning on creating a land tending day, a Traditional Ecological Knowledge (TEK) day. Should burning and land tending be on the same day? They go together in so many ways, like beans and rice. There is also the question of how much of our usual ancestral skills activities should we take out. Even without a land tending focus, the gathering is already a full week of ancestral learning. We will have to make room. We also truly believe that burning and tending are important ancestral skills, especially here in the western United States. Prescribed burning has always needed to be in the ancestral skills conversation. A part of this whole endeavor was to teach the participants that this is a vital ancestral skill to rematriate into our conversations. Definitely one that needs to move higher up the list.

We were aiming for this to be a community event. This was very important. A cultural community event. Everyone is invited, and everyone has a job. The elders will watch over all of us and let us know if we are doing it right. They will provide gentle, good-natured advice on how to do it a little bit better. The youth will light the fire. They will light it the old way without matches or a lighter. It will be lit using friction fire. Two teens, raised and mentored in the old ways, who are deeply connected to fire, plants, and the earth will be asked to spin the coal that will become flame.

After some discussion we decide that the two events will be separate, only because a fixed date will support planning the details for the TEK day. And with the fire we have to be fluid. We do not know when there will be a burn window. So we have to plan it in a way in which it can happen at a moment's notice. If it is on the same day, fine. If it is not, that is fine too. Even if no window opens at all, we have to be fine with that too.

We pick the first day of the gathering to be the TEK day. As the week gets closer, the weather forecast makes it look like Thursday will be a good day to "put some black on the land," to bring "good fire" back to the valley.

The TEK day comes and goes. It is a great day—as it turns out, a perfect way to begin the week. The people begin the week in good service to the land. The 250 participants are organized into six teams, with a knowledge holder or two guiding each team. Forest teams, meadow teams, creek teams, and invasive plant teams all collaborating in a coordinated plan, a plan designed for fire remediation and fire resilience.

In these times when there has not been enough fire on the ground, there is much work to be done to set the stage for a time when it is a regular thing again. To bring good fire to an unprepared site can easily lead to a bad fire, catastrophic fire, as we have seen in recent years. After a full day on the land, learning and sweating, many tired souls went to sleep with a smile on their faces. Grateful land spirits appeared in their dreams

The Fire

Thursday arrives. It is a cool morning. The air is mostly still. The humidity is good. Yes, this will be the day. An occasional gentle lake breeze brings the earthy smell of fresh tule growing along the shore. A juvenile osprey can be heard over the lake calling to its parents.

People begin to mill around, slowly coalescing into a group. We are waiting for the official burn boss, a person trained through the official, prescribed burn channels. This is one of the double paths we follow: the "legal" path, the paper rules path, the mainstream path, the settler path. He is the one that gets the permits, talks to the local authorities, and will be the one to make the call if there is a burn or not. The burn unit (the projected area to be burned) is given some last-minute attention. Duff is raked away from the old legacy grandmother oak. Even though she has seen a few fires in her day it feels good to offer some support and protection. Small seedling trees and bushes are earmarked for protection.

Some of the other burn leaders speak about the strategies for the day. What is going to happen? How will the fire move? Where can people plug in? The official burn boss arrives and shares some scientific knowledge of burning practices and results. We talk about how to work with the fire, how to make it do this, how to make it do that. We confirm the boundaries. We confirm what we find tolerable, such as the flame lengths we can feel safe with.

It is time to light the fire.

We offer the words
Words of gratitude
Words of blessing
The First Words
The Elder's words
We circle up for blue flame
We circle up for yellow flame
We circle up for red flame
We circle up for Fire

A long straight piece of wood or hard herbaceous plant stalk (such as mugwort or mullein), a flat piece of soft wood, and a bundle of tinder are all that is needed to light the fire. There are plant duos that work best together that bring the fire faster. The woody stalk is spun rapidly between two open palms as the bottom end of the stalk is pressed into a round notch in the flat piece of soft wood, into the hearth. When spun rapidly and consistently for a few minutes the hole blackens, as the friction creates heat. In the material being scraped away by the spinning, coals develop. The coals fall out of the notched opening into a soft bed of crushed mugwort leaves. After more spinning, a fragile coal pile is formed. When there is enough, they are carefully placed into a bundle of awaiting tinder. The bundle is picked up. The youth take turns blowing gently on the bundle. The coals are folded in deeper as smoke rises from the crisp, dry, organic plant material. The smoke gets thicker. The fire starters begin to blow more vigorously, until—POOF! A flame bursts from the bundle. The girl's face is lit up from the burst of flame. The gathered crowd offers a hardy but respectful cheer. There is pride for the fire starters, pride for continuing the old ancestral ways.

>Honor is given to the Fire starters
>Honor is given to the Fire
>Acknowledgement of its destructive power
>of its healing power
>of its history among the People
>The Fire is alive.
>We bring it to life
>We feed it
>We care for it.
>Prayers and blessings for the Fire
>For the trees
>For the animals
>For the land

This is the fire that will be used as the base fire for the whole burn. Torches will be lit from this fire. This is the fire we have talked over. This is the fire we prayed over. It is a special fire, and the flames from this fire will be shared while it is alive. It will heal the land. It will speak to the land. It will speak to the plants. It will speak to the animals. The smoke will also play its role, speaking to the plants and animals beyond the burn unit, while also smudging and fumigating the land.

The burn unit we have chosen for this day is a grove of pines, firs, cedars, and oaks. It is a small plot, but it will have a big impact—an impact on the land but also in the hearts and bodies of the participants and onlookers. The unit is in the middle of the camp for all to see, for all to smell, for all to feel.

We are burning the undercanopy. There is a lot of duff, and it is also a spot where there is an abundance of scotch broom, an aggressive immigrant plant that moves into disturbed areas. One of our objectives with this burn is seeking to control its spread and to bring balance to this space. We hope to give other native plants an opportunity to thrive here.

Many people from the village camp have gathered around to watch and or participate in the fire. There is a trepidation in the eyes of newcomers to the world of fire. Their previous learnings about the fear and menace of fire is strong. It is in their bones. Our lost relationship with fire leaves deep scars inside of us and in the world around us.

"What the heck? Is this okay?"

"What if it won't stop?"

"What about all the smoke?"

"Wait, fire is our friend?"

This small, seemingly insignificant fire will influence all who are here. It will ripple out to all of our larger circles. The newcomers stand back a bit but stay close, encouraged by the confidence and excitement of the others who are better acquainted with fire. The ones who have caught the bug are most eager to participate, the most eager to learn.

As people enter the world of fire, they find out that there is so much to learn, so much to know. Every fire is different. Every ecosystem needs a different approach. Every season has a different approach. Geography, climate, season, time of day, rainfall, plants, animals, and more are all part of the dialogue. This is a very important lesson: it takes training, and it takes experience. Fire is not a tool to be used lightly. For many it is the beginning of their journey into the world of fire.

The burn leaders move first. A shovel is used to pick up some coals from the main fire and move them to a pile of dry debris. The fire catches. Then the signal is given. Go! People step forward and begin. Children step forward, anxious to join, excited for the permission. As people move comfortably into their individual tasks, one wonders if we too, like the other wild creatures, are also hardwired for fire relations.

At first there is a flurry of action, at times appearing a little chaotic to an outsider. A local Cal Fire crew has arrived to check in on us, standard procedure. At first there is a look of surprise on their faces as they survey this scene. They have arrived right at a moment when there seems to be the most chaos, the most smoke. Little kids are rushing around in the smoke, adding pinecones to the mother fire and then fishing them out and carrying them to places that need fire. This is something new for these experienced firefighters. But as they continue to observe, they see there is order and purpose in the movements. Adults are gently guiding the children. "No, not there yet, over there." They see that the adults are in control, calm and purposeful. It is meant to be like this.

From afar, we can see children running in and out of the smoke, jumping over lines of fire. There is a relaxed demeanor among all, no fear, no anxiousness. Through the thick smoke it seems like the dark shadowy figures on the other side are the ancestors standing by, overseeing, breathing sighs of relief as good fire returns to the land.

At one point eventually the look of anxiousness and concern on the Cal Fire observers slowly, hesitantly turns to smiles. We see the two pathways struggling inside of them: the older, wiser path takes a foothold, and their shoulders relax. We wonder if this is an epiphany for them.

Then the initial excitement of the burn subsides. We settle into the phase of tending the fires burning throughout the unit. Wisps of smoke slowly rise from various piles, the hot spots. Children continue to jump over them. Elders rake unburned sections toward an awaiting flame. Tenders are moving fire around, spreading ash piles out, letting some areas burn out. There is a smell of burning pine needles wafting through our temporary village. It is a good smell, a smoky green tea smell. An expert can smell the difference between a good fire and a bad fire. Animals can too. They know when it is time to run. This is the "time not to run" smell. The good fire smell.

As the afternoon turns to evening, the easterlies slowly pick up. The fire is over. All the designated fuels have burned. Several mop-up monitors stand by, keeping an eye on the hot spots. A few of the children remain, poking through the ashes, also eyeing any potential hot spots. They will remember this day.

An eagle swoops down off a snag left from an old wildfire and skims across the lake, snatching a salmon deftly out of the water. Butterflies on the shore mirror the eagles' movements as they swoop in and out of the lupines.

California is a big state. It took a monumental effort to tend to so much land on a regular basis. To revive the land, to heal the land, prescribed fire needs to make a huge comeback. We see this as a practice to be done all the time, everywhere. It will take all of us, in collaboration, in community. That's how it was in the old days: every tribe, every tribelet, every village doing their work to tend to the land. And it was a year-round endeavor. Each season has its tasks. Literally millions of people were tending to our home on a daily basis.

One of the key ingredients to this healing, rematriating process is to bring the old ways, the cultural ways, back into the process. For us to get back to that place where we were in pre-colonized times, relatively safe from huge catastrophic fire, we all must be educated on the practices of our Indigenous ancestors. It is good to know the facts, the scientific side of fire and controlled burning, but it is the cultural knowledge and practices that will hold it together, the glue that will sustain the practice. These two knowledge bases need to work together.

One cultural aspect of prescribed burning that needs reintegration is the community participation in planning and implementation. There is so much discussion that can take place in the planning of these burns. Here we use another one of our ancestral skills, dialogue. All the interested parties are in the conversation. The elders, the youth, the grandmothers, the basket weavers, the herbalists, the hunters, and more all have a say in when, where, and how fire is put on the land. Deals are made. We negotiate for the deer, for the bear, for the sedge roots, for the oak trees. It is a continuous, evolving dialogue of the land. In the next season we do it all over again.

Our biggest dream is that these children will be present for the next season's dialogue. We will remember all the children who participated. This is the beginning of their training. We are training up our future burn bosses. We will need all of them.

Epilogue

They say that when we use fire we are creating space for the indigenous ecosystems to return. We are pushing back on the aggressive plants, both immigrant and native. When we make space available, the less aggressive herbaceous plants have an opportunity to thrive. So often the seeds are just there, waiting, in the seed bank, in the soil, waiting for their chance to come back again. All that is needed is for a cold fire (good fire), just burning off the top layer, to move quickly through, to scarify the seed, to open the canopy a little, to create some open ground, to add nutrients to the soil, to free up more water. California native plants have evolved in this system. It works for them. That is why we are always saying that fire brings back the natives in this way.

Recently the total truth of that statement was impressed on us at an even deeper level. There have been catastrophic fires here in California in recent years. (It is 2022 as of this writing.) In several of the really big record-breaking fires, some of the blame for that has gone to PG&E. They were sued, and they had to declare bankruptcy. As a part of the resolution of those proceedings, they had to give up some of their vast landholdings. Most of those grants went to environmental and conservation organizations, supposedly those groups that know how to manage large tracts of wildland. In the Sierras, an Indigenous group was formed, and they asked to be put on the list and were denied—they were told they couldn't manage large tracts of land! The group took the issue to court and won. They received thousands of acres. So here we have it: fire has returned natives back to the land, human natives. The Maidus got land back!

From the Maidu Summit Consortium website:

> Tásmam Koyóm is the Maidu name for the valley that is located in Plumas County, California. This valley was an important Maidu population center within the traditional homeland of the Mountain Maidu for many generations. . . .

Tásmam Koyóm was returned to the Maidu Summit Consortium on September 20, 2019[;] this was a historical moment in MSC history. Pacific Gas & Electric transferred 2,325 acres of land back to the Mountain Maidu people. The consortium started in 2003 for this purpose[;] it's been a long wait for the Maidu people, volunteers, and the vision that Farrell Cunningham and the MSC board members had to finally bring our land back home to our people.

The Maidu began doing prescribed burns right away.

Afterword

As this book approached its printing date, fires erupted in the southern half of our lovely California. In early January, smoke from the Palisades Fire billowed along the Pacific Ocean as the flames began consuming 23,000 acres and 1,200 homes there. Later that same day, the Eaton Fire ignited about thirty-five miles east, in the foothills of the San Gabriel Mountains; the Santa Ana winds charged through the hills with 100 mph gusts that evening, shoving the blaze toward rows of homes. While smaller in scale at roughly 14,000 acres, the Eaton Fire destroyed over 9,000 homes and other structures. A total of twenty-nine lives were claimed in these fires.

Smoke hovered across the region for weeks and in Pasadena, at least—where I (Dani) happened to be for most of January—ash coated cars, parking lots, and sidewalks like a light dusting of snow. Massive piles of palm fronds, tree branches, and other yard waste sat in heaps along every block, evidence of the ruthless winds that tore through Southern California on those first nights. The fear and grief, the smoke-headaches and hazy skies, it was all too familiar.

It was impossible not to think of Octavia Butler, who grew up in Pasadena and is buried at the Mountain View Cemetery in Altadena (which was partially burned), during the weeks of fire. Her 1993 book, *Parable of the Sower*, is set in a post-apocalyptic Southern California between 2024 and 2027 during a time when the impacts of climate change and poverty are dire, and capitalism has caused societal collapse. Police and other officials had become corrupt and untrustworthy during emergencies.

The state of our country in early 2025 isn't wildly different than what Butler described in her book. During the Eaton Fire, a fascist president was inaugurated for a second time and the richest man on earth gave a Nazi salute at the inauguration. Reports spread throughout Southern California that ICE (Immigration and Customs Enforcement) had begun raids on undocumented communities in neighboring Kern County, one action of what would become many deportation attempts as promised by the current administration. The president yammered to television news crews about turning on some great mythical faucet in Northern California to send water flowing to the parched hillsides and streets of the south. He did, in fact, order two billion gallons of water to be released from two reservoirs; water that should have been stored for use in the hot, dry summer but was instead wasted as it gushed into Central California, far from the fires. Throughout the smoldering neighborhoods in Altadena, members of the National Guard and local law enforcement loafed around checking their cell phones, drinking coffee, and blocking roads with weapons in plain view. A third fire sparked near a detention center and authorities refused to follow evacuation orders to get the inmates to safety as the flames inched closer to the facility. I stayed glued to the news, waiting for someone to get them out, angry that the 4,700 men inside were treated as disposable (note: firefighters were able to prevent a disaster).

It felt very much like an Octavia Butler prophecy come to life, except it wasn't a work of futuristic speculative fiction. We can't just turn the pages and reshelve this book. The ever-increasing climate catastrophes are here. They are written into our lives.

Yet despite the gloom and overwhelm about climate change catching up to us and regularly destroying homes and lives and communities, there was something else I witnessed while driving to check if my sister's favorite local grocery store was still standing (it was not) or walking to volunteer at a distribution site. In stark contrast from the heavily armed military members, residents gathered in driveways, in front yards, in parking lots and on sidewalks where they sorted clothing, food, and water to distribute to fire survivors. There were plastic tables with warm meals available. Boxes of masks and shoes. There were individuals and families at intersections handing out bottles of water. None of them were government agencies; just regular, everyday folks stepping in to help.

One particularly impressive place, the Pasadena Job Center, an organization that connects small businesses and homeowners with skilled day laborers, coordinated an extremely organized distribution site with hundreds of cars rolling through daily to receive food, water, toiletries, clothing, and even pet food, ice cream bars, and other items. At least two different groups served free warm meals each day. Workers from the center even cleared the piles of debris, which had sat there for weeks—from the nearby neighborhoods as high winds were predicted yet again about two weeks into the fire.

Mask Bloc Los Angeles, one of a large number of international independent mutual aid groups that provides high-quality masks to their communities for free, supplied roughly 300,000 masks around Los Angeles County before the end of January. Posts on social media reported that government officials from the City of Los Angeles were so unprepared, they even reached out to the group to ask for masks.

It confirmed, yet again, something I've long believed to be true: we don't need to wait for "officials" to show up. We have the power and knowledge and ability to care for one another like this at the drop of a hat. Now just imagine if we were all even *more* prepared, if we had community networks in place to step in for every possible disaster we might be confronted with. Imagine if we took some of that power that so many look to the government for and shared it with each other.

There is a scene in *Parable of the Sower* when the protagonist, Lauren, and her group have reached Northern California and are walking Highway 20 through Lake County; both sides of the highway are engulfed in flames. Lauren can feel the heat on her skin as she maneuvers down the center of the two-lane road curving through the burning landscape. I was rereading the book in 2015 when I found myself heading down that same highway on my way home from Tehama to Sonoma County during the first day of the Valley Fire. The smoke was thick and rushed toward me in heavy dark clouds, and visibility came and went as the wind gusts pushed the smoke back and forth through the hills of oaks. I was not prepared. My kids and I grabbed whatever scarves and sweaters were in my car to cover our faces. Our eyes burned. Roads were closed and chaotic. It was terrifying.

But in the middle of the chaos there was something I witnessed that I will never forget: families gathered on the side of rural roads with empty horse trailers and homemade signs offering safe places to take pets and livestock for evacuees. As I sat in traffic, waiting to be directed through a detour, I witnessed people stopping to give out water, to hug weeping people arriving with their horses and goats. Just everyday people doing what they could to care for their community, to ease the fear and grief just a little bit. It was September and the sun was setting, and despite the smoke, that hazy orange sunset and community care was one of the most beautiful things I've ever witnessed. I will never ever forget that scene as long as I live.

It is so easy to be all-consumed by the rage and heartbreak, to lose sight of the fact that every time a climate catastrophe hits, we have an opportunity to build the world we all deserve to live in. So many times in recent years I've returned to this line in *Parable of the Sower*: "That's all anybody can do right now. Live. Hold out. Survive. I don't know whether good times are coming back again. But I know that won't matter if we don't survive these times."

I do believe that we indeed owe it to each other to survive these times. Or at least to try. Both Margaret and I hope the stories in this book have offered routes toward that possibility of creating a better world. We can only get there if we make it through these times to create a new world by and for ourselves.

Xo
Dani & Margaret

Acknowledgments

First and foremost, we offer gratitude and respect to the many Indigenous people of California. Next, this book would not be possible without the many inspiring folks featured in these pages who contribute their insights, efforts and commitments to making our communities and this world a better place. Thank you for your time and belief in this project, and for the mutual aid you bring to this world. A huge bow of gratitude (and high five) to Zach and the rest of the folks at AK Press for believing in and encouraging us along the way, despite fires and a spiraling world. Thank you to Saif Azzuz for the use of your beautiful art on our cover. Thank you to *Yes! Magazine* for publishing variations of the piece Dani included in this anthology. My (Dani's) whole heart to my kids, Enfys, and Ava for listening to me talk about this book, and fires, and mutual aid for more years than I can count. Sorry for traumatizing you. Thank you to *Feather Publishing* for support and encouragement during and after the fire and for publishing earlier versions of Margaret's work. A giant thank you to husband Ryan, and children Paloma and Diego for navigating our post-apocalyptic world together in our new homeland, and to sweet Kaytlyn—who lost her childhood home to the Dixie—thank you for keeping the rest of us grounded. And to the memory of Farrell Cunningham, who taught me an exquisite appreciation for his Maidu homeland—I remain always in his debt.

Contributors

Saif Azzuz: is a Libyan-Yurok artist living in Northern California. More about Azzuz is here: saifazzuz.com.

Freelance writer **Jane Braxton Little** focuses on science and natural resources in stories published in the *Atlantic, Bay Nature, National Geographic, The Nation, Scientific American,* and other magazines. In 2023 she helped launch *The Plumas Sun,* an online news platform serving climate-disaster communities in Plumas County, California.

Dani Burlison (she/her) is the creator/editor of "All of Me: Stories of Love, Anger and the Female Body" (PM Press, 2019), the author of "Some Places Worth Leaving" (Tolsun Books, 2020) and "Dendrophilia and Other Social Taboos: True Stories," a collection of essays that first appeared in her *McSweeney's Internet Tendency* column of the same name. She's been a staff writer at a Bay Area alt-weekly and a regular contributor at *Yes! Magazine, Chicago Tribune, KQED Arts,* and elsewhere. Her journalism, fiction and personal essays can also be found at *Ms. Magazine, WIRED, Earth Island Journal, The Rumpus, Portland Review, Hip Mama Magazine,* and in various anthologies and zines. Dani grew up in a rural, working-poor family and community in rural Tehama County and is a single mom to two magnificent young adults. You can find out more at daniburlison.com.

Beatrice Camacho is a first-generation Chicana raised in Sonoma County to low-income, working-class parents who immigrated from Northern Mexico. As a lifelong renter, growing up in Section 8 Housing, she knows the importance of dignified and affordable housing. Beatrice holds a B.S. in Business Management and is trained in Restorative Justice and Restorative Practices. An Organizer with the North Bay Organizing Project since 2018, she organized the Sonoma County Tenants Union, dispersed over $5 million in Emergency Rental Assistance in 2021, became the first Director of Undocufund in January 2022, and co-founded the statewide California UndocuFund Network in September 2022.

Lasara Firefox Allen, MSW, (they/them/Mx) is the author of "Jailbreaking the Goddess" and "Sexy Witch," as well as chapbooks "The Pussy Poems" and, as contributor and editor, "Disjointed." Currently enjoying writing micro memoir in addition to their more established nonfiction genre, their work has been recently published at *Sledgehammer Lit, LiteraryKitchen.net, Spooky Gaze, Tangled Locks Journal, Mountain Bluebird Magazine*, and in the *Guilt Scar* zine. Lasara is a Witch, a nonprofit Executive Director, a menopause and life coach, and a co-conspirator for our collective liberation. Learn more at: https://linktr.ee/Lasarafirefoxallen.

Margaret Elysia Garcia (she/her) is the author of the poetry collection *the daughterland* (El Martillo Press, 2023), the short story collection *Graft* (Tolsun Books, 2022), and the poetry chapbook *Burn Scars* (Lit Kit Collective, 2022). She was a reporter for *Feather Publishing* from 2017 until the paper folded in 2023 and now writes for its magazine *High Country Life*. Her short stories, essays, and poetry can be found at various journals in print and online including

Contributors

Querencia Press, Latinos in Literature, Huizache Journal, Catamaran, Hip Mama, Somos en Escrito and *Raven's Perch*. Originally from Los Angeles, Margaret and her family made her home in Plumas County for two decades until the Dixie Fire rendered them climate change survivors. Readers can find out more at www.margaretelysiagarcia.com.

Kailea Loften is a Black American, citizen of Liard First Nation and member of the Tsesk'ye clan from Nalokoteen (end of the ridge nation) of the Tahltan Nation. For the last decade she has worked at the intersections of climate justice, spiritual ecology and independent publishing. Since 2019 she has been the co-editor of Community Publisher *Loam*. She has guided climate change policy with an emphasis on Indigenous rights on local, national and international levels, previously serving as a Climate Commissioner for the City of Petaluma, and as the Climate Justice Organizer and Community Publisher for NDN Collective.

Zeke Lunder is a wildfire analyst from the Lassen County town of Westwood. His website the-lookout.org is a valuable public service tool for anyone interested in public data that is then analyzed for the general public in regards to fire and water. Fire information of the Sierra Nevada has proven invaluable to residents of the Lost Sierra region.

Manjula Martin is the author of *The Last Fire Season: A Personal and Pyronatural History*, and coauthor, with her father, Orin Martin, of *Fruit Trees for Every Garden*, which won the 2020 American Horticultural Society Book Award. Her nonfiction has appeared in *The New Yorker, Virginia Quarterly Review, The Cut, Pacific Standard, Modern Farmer,* and *Hazlitt*. She edited the anthology *Scratch: Writers, Money, and the Art of Making a Living*; was managing editor of Francis Ford Coppola's literary magazine, *Zoetrope: All-Story*; and has worked in varied editorial capacities in the nonprofit and publishing sectors. She lives in Sonoma County, California.

Margo Robbins is the co-founder and Executive Director of the Cultural Fire Management Council (CFMC). She is one of the key planners and organizers of the Cultural Burn Training Exchange (TREX) that takes place on the Yurok Reservation twice a year. She is also a co-lead and advisor for the Indigenous People's Burn Network. Margo comes from the traditional Yurok village of Morek, and is an enrolled member of the Yurok Tribe. She gathers and prepares traditional food and medicine, is a basket weaver and regalia maker. She is the Indian Education Director for the Klamath-Trinity Joint Unified School district, a mom, and a grandma.

Amy Elizabeth Robinson is a poet, writer, historian, mother, and many other things. Upon completing her Ph.D. in the history of British imperialism in 2005, she decided to not pursue an academic career and instead moved to Moman's Rill, a collectively-owned community on ancestral Wappo lands in the eastern mountains of Sonoma County, California. Since then, she has tended the lives of her family, community, and forest, been involved in several activist movements, trained as a community leader at Flower Mountain Zen, and has written and published world history curriculum content as well as poetry and essays. In September 2020, however, most of Monan's Rill burned in the Glass Fire. Now she is rebuilding in community. Amy's work has appeared in *Literary Hub*, *Literary Mama*, *New Verse News*, *West Marin Review*, *West Trestle Review*, *Rattle*'s Poets Respond program, and elsewhere. She also blogs at www.turningplanet.org.

Brandon Smith is the senior advisor and co-founder of the Forestry and Fire Recruitment Program. Smith grew up in Altadena, California and has a Bachelor of Arts in African American and Liberal Studies from the University of California, Berkeley. While incarcerated later in life, he was approached to join a Conservation ("Fire") Camp. Brandon met his co-founder, Royal Ramey, at the Bautista Adult Conservation Fire Camp in Riverside County. Smith was released in March 2014 after his sentence was reduced because of his work on a fire crew. After release, Brandon spent over eighteen months applying to fire stations before receiving his first professional assignment to fight the Lake Fire in 2015. Shortly after, Brandon was hired by the US Forest Service.

Hiya Swanhuyser is a writer from Northern California. She is a founding member of Substrate Arts, a co-operative online magazine covering Bay Area creative scenes. Her work has appeared in *the Believer, KQED Arts, Fandor's Keyframe, Otis Nebula, Petals & Bones, the Press Democrat*, the *North Bay Bohemian, Mission Local, San Francisco Magazine, SF Weekly*, and *7x7*. She grew up on a family farm in the counterculture, and is interested in public drinking fountains and Freedom Riders. She is at work on a book about the Montgomery Block, a lost bohemian refuge of San Francisco's North Beach. In her free time she volunteers at the Phoenix Theater, hosts live music in her back yard, and stands in public spaces calling for ceasefire, antiracism, and land back.

For over thirty-five years, **Sue Weber** has worked with global NGOs with specific emphasis on disaster relief work. She is the founder of Indian Valley Academy under Plumas Charter School and the founder of Plumas Strong, a non-profit dedicated to enhancing the lives of the people of Indian Valley with specific emphasis on the youth. She was the co-chair, then chair of the Dixie Fire Collaborative, which is dedicated to the recovery of Plumas and Lassen counties' communities impacted by the Dixie Fire. She is currently the executive director of KIDmob and chair of the Indian Valley Park and Recreation Department.

Redbird Willie: After graduating from UC Berkeley with a degree in native studies Edward (Redbird) Willie has continued to enrich his personal education, fueled by his desire to uncover and rekindle the cultural earth-based knowledge of California Indians. He teaches traditional ecological knowledge (TEK), fire ecology, permaculture, and ancestral skills to people of all ages. He has, in recent years, been a core organizer of the annual Buckeye Gathering, an ancestral skills gathering. He has been an adjunct instructor for Weaving Earth, an organization dedicated to reconnecting humans to each other and to the earth. He is currently working for The Cultural Conservancy as the land steward for Heron Shadow. Also an artist (drawing painting and sculpture), he has recently illustrated a children's book, *The Adventures of Two Coast Miwok Children*, and has been featured in *News from Native California*.